Effects of GINKGO BILOBA EXTRACT on ORGANIC CEREBRAL IMPAIRMENT

Effects of GINKGO BILOBA EXTRACT on ORGANIC CEREBRAL IMPAIRMENT

A selection of thirteen papers presented at the
International Symposium held in
New York: 22nd September 1984

Presidents:
Professor A. Agnoli (Italy)
Professor J. R. Rapin (France)
Professor V. Scapagnini (Italy)
Professor W. V. Weitbrecht (West Germany)

John Libbey
EUROTEXT
London · Paris

British Library Cataloguing in Publication Date

The Effect of ginkgo biloba extract on organic cerebral impairment: proceedings of the international symposium held in New York, September 1984.

1. Brain Damage — Treatment 2. Ginkgo
I. Agnoli, A.
616.8 RC387.5

ISBN 0-86196 064 5

Published by John Libbey & Company Ltd
80/84 Bondway, London SW8 1SF, England.
(01) 582 5266

John Libbey Eurotext Ltd
6 rue Blanche, 92120 Montrouge, France. (1) 657.38.38

Printed in Great Britain by Whitstable Litho Ltd, Whitstable, Kent

Contents

Introduction

Statistical data which have just been published shown that the average life-span is 73-75 years in industrialized countries. The increase in life-span observed during the past decades is unfortunately linked with an increase in the incidence of senile dementia and risks of cerebral ischemic insult.

A policy of prevention should absolutely be set up to prevent these disorders ; amongst the available means are the so-called "vasoactive drugs". Their efficacy, either curative or preventive, is still debated. This justifies the amount of research made by the pharmaceutical industry to answer the requirements for official authorities.

The extract of Ginkgo biloba belongs to this therapeutic group and the pharmacological as well as the clinical works presented during this symposium show objectively some aspects of its efficacy. It is a complex product with multiple pharmacological properties with the possibility of additive effect, synergy or antagonism.

The first proposed mechanism is an anti-edema effect which is shown whatever the type of experimental edema : toxic, traumatic or ischemic. This beneficial effect is clinically found again during vascular attacks. In this case Ginkgo biloba extract induces a re-adjustment or a recovery of the blood flow and functional activity.

The mechanisms involved are numerous and are related to hemorheology (Ginkgo biloba extract standardizes the abnormal blood rheology), to radical scavenging effects and to the restoration of membrane integrity. As well as its effect on ischemic disturbances Ginkgo biloba extract has been proven to be active in deficits due to ageing and especially in primary degeneration dementia.

This effect must be confirmed in all cases of predemential syndrome to conform with our concept of prevention. The active mechanisms can be linked with vascular effects in areas where degeneration is ongoing, and are involved with the aminergic and the cholinergic sytems : Ginkgo biloba extract at least induces an increase in the number of acetylcholine receptors in aged animals.

At the present time, there is no sufficient evidence to advocate the use of this extract in the treatment of Alzheimer disease but its use can be justified in patients presenting with the first symptoms of dementia. Although the problems in cerebral ageing are different from that of dementia, they become each year more and more worrying. The causes of the problems are unknown but at this point it is useful to develop the research of suitable treatment and to base this research mainly on preventive therapy.

These treatments must be effective in many areas and this explains the necessity of using several active principles. The polymorphic aspects of cerebrovascular diseases lead to the same approach to its treatment. We now have to build a rational polymedication which can involve several drugs or a complex natural extract such as Ginkgo biloba extract.

This symposium has indicated the way to progress in the treatment of organic impairment ; presentations and discussions have shown the positive effect of multifactorial approaches.

Professor J.R. RAPIN

I.N.R.P.V.C. - Hôpital Bicêtre, 78 rue du Général Leclerc, 94270 Le Kremlin Bicêtre, France.

GINKGO BILOBA EXTRACT

Experimental Studies

Effects of Ginkgo Biloba Extract on Organic Cerebral Impairment. A. Agnoli, J. R. Rapin, V. Scapagnini, W. V. Weitbrecht. John Libbey Eurotext Ltd

Effects of Ginkgo biloba extract on cerebral metabolic processes

S. S. Chatterjee

Department of Pharmacology Dr. W. Schwabe, Arzneimittel, Karlsruhe, West Germany

ABSTRACT

The aim of the present communication is to summarize some of the pharmacological studies carried out with the extract of Ginkgo biloba (GBE) and discuss our current concept of GBE as a multiple therapy principle.
We have demonstrated that GBE possesses vascular-tone regulating properties, an antihypoxic effect and experimental cerebral oedema and neurotoxicity inhibitory actions. Known mechanisms of action of GBE involving radical scavenging, prostacycline synthesis stimulating, modulating effect on the cerebral energy metabolism cannot explain the global efficacy of the extract observed under therapeutic conditions. The clinical beneficial effects of the extract are due to the combination of its various protective, curative and modulating properties against the pathological processes.

KEYWORDS

Ginkgo biloba extract - cerebral metabolic processes - pharmacological studies review -

INTRODUCTION

The leaves of Ginkgo biloba (Maidenhair Tree) contain a number of flavone glycoside esters together with some Ginkgo biloba specific terpene lactones: Bilobalides and Ginkgolides A,B and C. An extract standardized on specific components is now being widely used in various European countries for the treatment of peripheral and cerebral vascular diseases. The efficacy and usefullness of this standardized extract has been demonstrated in various clinical studies (1; for review see 2). In the present communication the pharmacological studies carried out with the extract in our laboratories will be summarized and our current concept on the possible mode of action of the drug will be discussed.

VASCULAR TONE REGULATING PROPERTIES OF GBE :

The extract of Ginkgo biloba was introduced in the therapy of vascular diseases 1975 and this was based on the observations that the extract was able to lower the vascular resistance under some experimental conditions (3). Soon, however, the observed clinical activity profile of the extract and the absence of side effects like "Steal-phenomenon" etc..., which are commonly observed with vasodilating substances, made it clear that the beneficial effects of the drug cannot be completely explained only on the basis of its vascular tone regulating properties. In addition pharmacological studies carried out in our laboratories have demonstrated that the vascular effects of the extract cannot be defined in terms of classical vasodilation, as it does not relax the isolated smooth muscle preparations (4). Our experimental findings suggest that the vascular resistance lowering effects are due to some of its metabolic effects. In this respect, prostacycline synthesis stimulating properties of Ginkgo biloba may be of some importance and, therefore, would be discussed later in some details. It is, however, important to mention here that the studies conducted by others, specially those by Defeudis et al. (9), have been able to demonstrate that the vascular tone regulating actions of the extract are complex and probably dependant on the pathological state and the type of vasculature involved.

ANTIHYPOXIE ACTIVITY AND RADICAL SCAVENGING ACTION OF GBE :

Now that, during the mid 70's it was getting clear that some other activity of the extract in addition to its vascular effects must be responsible for its clinical beneficial effects, we started looking for it. Soon, it was realized that the extract, unlike many other known vasodilators, was able to prolong the survival time of animals under hypoxia (Fig.6). This antihypoxic activity of the extract could be demonstrated under various experimental conditions. Efforts to clarify the mechanism of this antihypoxic effect led us to the discovery of one of its biochemical properties, a radical scavenging activity. Initially we reached this conclusion only through comparative pharmacological studies conducted with the extract and other known substances. The only other drugs or agents which were found to prolong the survival time of the animals under hypoxic conditions were barbiturates, benzodiazepines and a large group of substances of different chemical structures, possessing radical scavenging properties in common. These observations were in agreement with the proposal made by Demopolos and others (7) who suggested that the antihypoxic and antiischemic properties of barbiturates, atleast under experimental conditions, are not due to their sedative or hypnotic properties, but rather to their radical scavenging actions.

Further, we could demonstrate that the extract, like barbiturates and other radical scavengers was able to inhibit radical induced toxicological processes under *in vivo* conditions. The experimental models chosen for these studies were the alloxan induced diabetes and the adriamycin toxicity. It has repeatedly been reported that these two toxicological processes are triggered by radical induced reactions initiated by the toxicons. It can be seen in Table I and fig. 1 that the extract like other radical scavengers is effective in inhibiting the toxicities of alloxan and adriamycin only when the animals were treated with the agents in a protective manner i.e. when the scavengers or the extract were given before the administration of the toxicons.

Table I : Agents which show anti-hypoxic activity and afford protection against alloxan diabetes.

1. Known radical - Scavengers
   - Ethanol, Methanol, Propanol, Butanol
   - Dimethyl - Sulfoxide
   - 1-Phenyl - 3 - (2 - Thiazolyl) - Thiourea
   - N' N' - Dimethyl - urea
   - Benzoic acid
   - Ascorbic acid
   - Barbituric acid

2. CNS - Active Substances
   - Diphenylhydantoin
   - Diazepam, Oxazepam and other Benzodiazepines
   - Pentobarbital, Hexobarbital and other Barbiturates

3. Anti - Hypoxic agent
   - Extract of Ginkgo biloba

Fig. 1 : Protection of Adriamycin toxicity by the extract of Ginkgo biloba. Note that the protective effect is observable only when the extract is given before the Adriamycin administration.

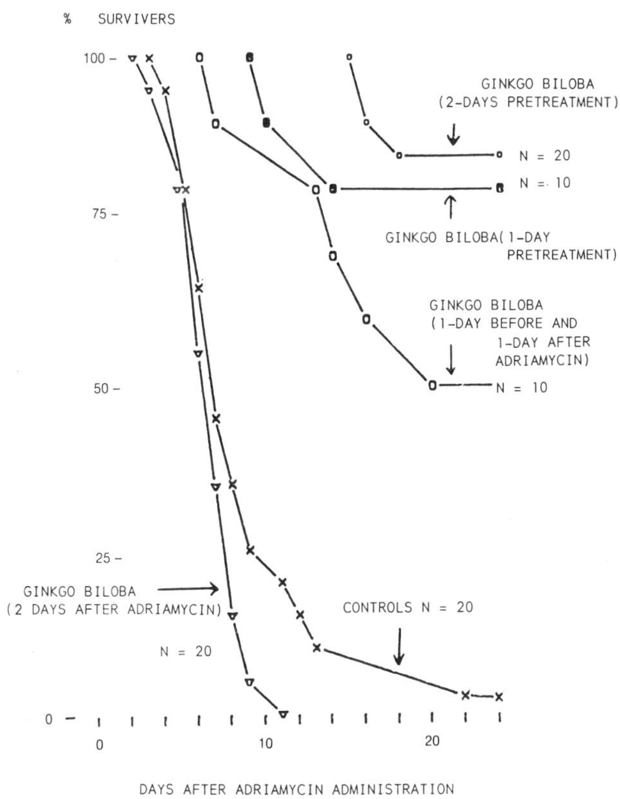

These experimental evidences suggested that some of thebeneficial action of the Ginkgo biloba extract could be due to its radical scavenging property.

This property has been since confirmed by several authors using different experimental conditions (8,9,10). Some of the results obtained in our laboratories (11) are shown in Fig.2.- The radical induced lipid peroxidation in brain homogenates of rats is inhibited by the extract in very low doses (10-20 µg/ml) in a dose dependant manner. Treatment of the animals with the extract reduced the lipid peroxide levels in the brain.

Fig.2 : Inhibition of lipid paroxidation by Ginkgo biloba extract under *in vitro* (A, rat brain homogenates) and *in vivo* (B, mouse brain) conditions.

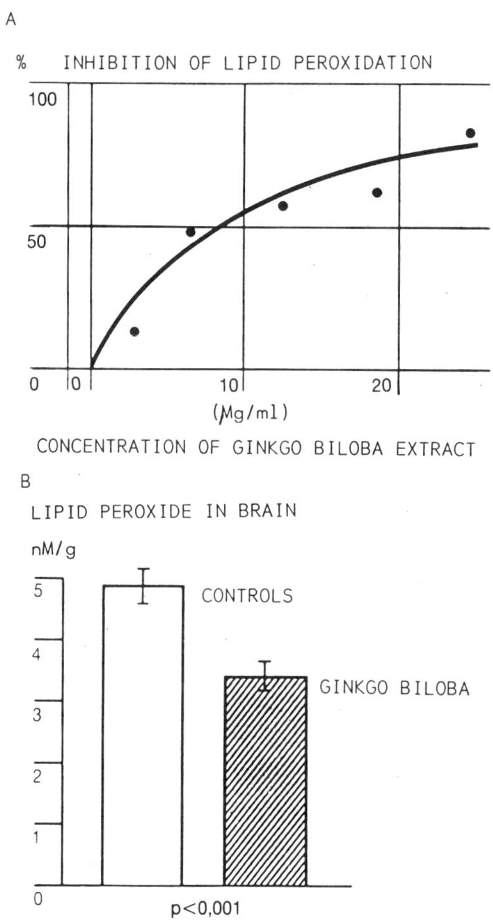

PROSTACYCLINE SYNTHESIS AND GBE :

The consequences of this radical scavenging property of Ginkgo biloba for the therapy of ischemic diseases are far-reaching and is beyond the scope of this review (for more details s.11) and therefore would not be discussed in details here. Our attention was drawn to the observations made by different authors (f.Rev.s.12) describing the regulating and modulating effects of radicals and radical induced reactions on the arachidonic acid metabolism (Fig. 3).

Fig.3 : Some key steps involved in the arachidonic acid metabolism are depicted. Free radicals are involved at several steps and modulation of these radicals or radical induced reactions drastically influence the product ratios of the metabolic pathway.

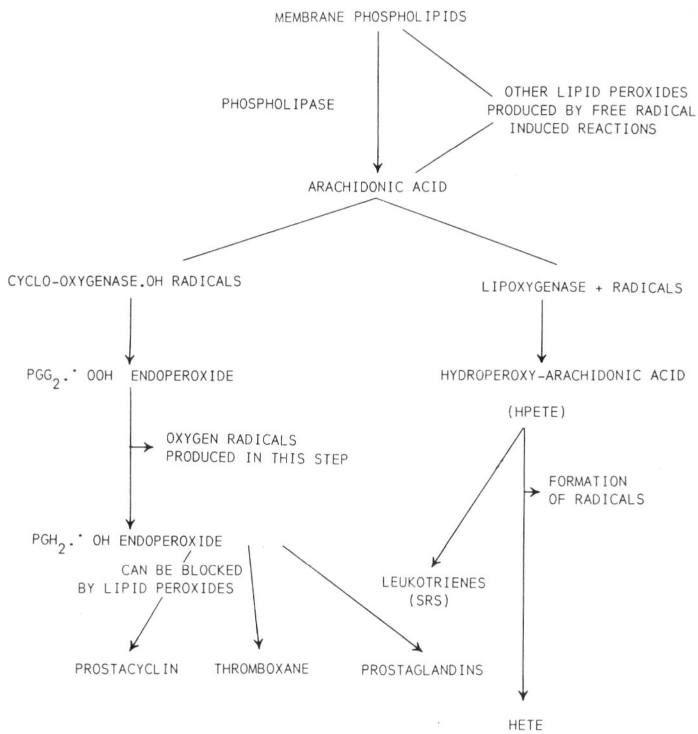

The importance of the different metabolites of the arachidonic acid pathway : prostaglandins, thromboxane, prostacycline and leukotrienes in the pathogenesis of vascular ischemic diseases and other disorders are now well known. In this connection it is important to note that the lipid peroxides can disturb the balance between thromboxane and prostacycline - which is of vital importance in the regulation of thrombogenesis and other circulatory disturbances - by inhibiting the prostacyclin synthesis (13). If the extract of Ginkgo biloba inhibits lipid peroxidation, then it would be logical to expect that as a consequence the prostacycline synthesis must be stimulated. Indeed, it could be demonstrated that incubation of vascular tissues with the extract in very low doses potentiates the prostacycline synthesis in a dose dependant manner (Fig.4). These observations from our laboratories have been confirmed by Herman et (14) using specific RIA methods for the determination of prostacycline degradation products.

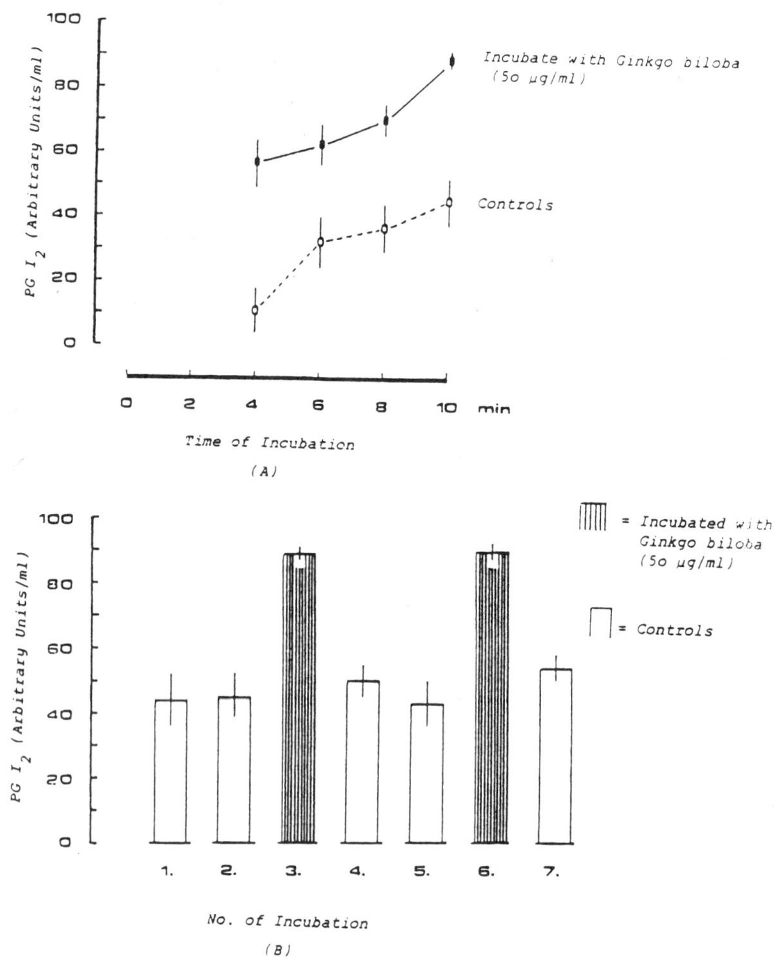

<u>Fig</u>. 4 : Effect of Ginkgo biloba extract on the prostacyclin synthesis in the rat aorta strips incubated in vitro.
(A) The aorta strips were incubated for different periods.
(B) The aorta strips were incubated for lo min, each time with intermittent thorough washings. All incubations were started after repeated washing of the strips with the incubating medium.

CEREBRAL ENERGY METABOLISM: OXYDATIVE PHOSPHORYLATION AND GBE :

At this stade during 1979 - 1980 we were quite satisfied with our findings and thought that the global mode of action of the extract could be explained on the basis of the scheme shown in Fig.5.

Fig.5 : Scheme explaining the mode of action of Ginkgo biloba extract on the basis of its radical scavenging properties. The extract is known to inhibit the processes marked by (---).

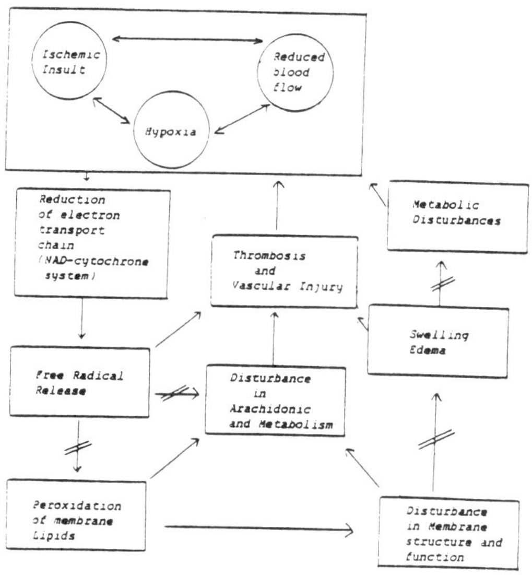

It soon, however, became clear that there were other clinically important properties of the extract that could not be explained by its radical scavenging property. It was reported by Krieglstein's group (15) that the antihypoxic activity of the extract was associated with changes in the cerebral energy metabolism (Fig. 6), which could be responsible for this activity.

Fig.6 : Effect of Ginkgo biloba extract on cerebral energy metabolism and an survival time of rats subjected to hypotoxic hypoxia (data from prof. Krieglstein).

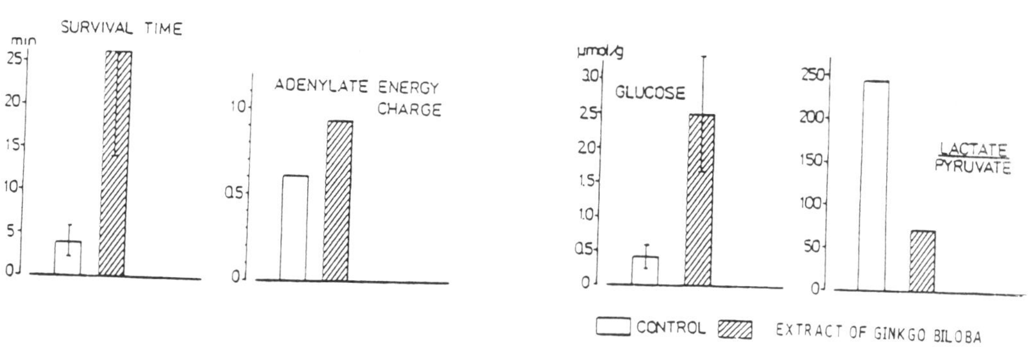

Similarly Rapin et al. (16) demonstrated that the cerebral energy metabolism regulating properties of the extract under ischemic and hypoxic conditions could be responsible for its beneficial effects. In addition the experiments conducted by Cahn and his colleagues (17,18) demonstrate that an experimental oedema induced by triethyltin or trauma could be cured by oral treatment of the animals with the extract. Since, till now, no conclusive evidence for the involvement of radical induced processes in these pathological models are available and because it has been categorically demonstrated that the radical scavengers or antioxidants are not active in these models, it was considered possible that the extract possesses additional properties other than radical scavenging which enables it to modulate the pathological disturbances in oxidative phosphorylation or cerebral energy metabolism. In order to get some information on this additional property a project have been initiated in our laboratories in which the effects of the extract on uncoupled oxidative phosphorylation or on cerebral energy metabolism are studied.

Thriethyltin is known since long to be an uncoupler of oxidative phosphorylation and, therefore, it seemed logical to study the effects of the extract on triethyltin toxicity. This led us to some surprising findings which, for the sake of brevity, are not possible to be described here in details. In short, it was soon realized that the cerebral oedema, which could be cured by treatment with the extract of Ginkgo biloba, was not the primary toxic effect of the uncoupler, but rather demyelination and disturbances in nerve conduction are the primary signs of its toxicity. In addition it could be demonstrated that preventive or curative treatment of the triethyltin intoxicated animals with the extract prevents or cures the neurotoxic signs of intoxication (Fig.7).

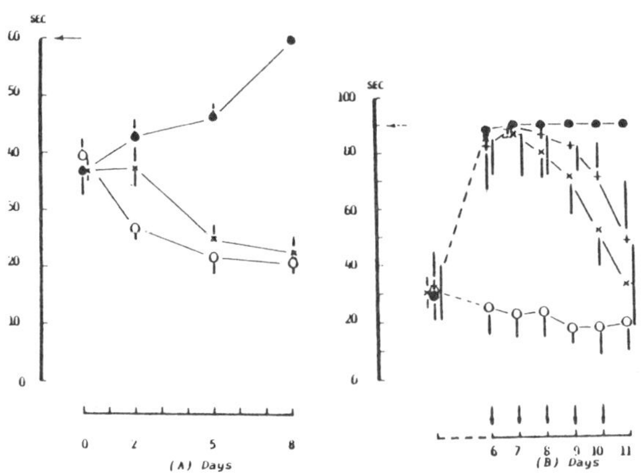

Fig. 7 : Protective and curative effects of the extract of Ginkgo biloba (EGB) on triethyltin (TET) toxicity. The vertical axis represents the reaction time of the animals (in secs.) on hot plate test, which is prolonged after TET intoxication (control animals without intoxication = 0-0 TET intoxicated animals = .-. = TET intoxicated animals treated with EGB (100 mg/kg/day and 50 mg/kg/day x-x and +-+ respectively). In the experiments shown in (A) the rats were intoxicated with TET by incorporating it with drinking water (0,002%) and the treatment of the intoxicated animals were starded simultaneously by oral application (100 mg/kg/day during 8 days). In the experiments shown in (B) the rats were intoxicated with TET by oral application (1,65 mg/kg/day for 5 days) and then divided into 3 groups. One group receided daily, from day 6 onwards, oral application of water (10 ml/kg) and the other two groups received EGB (100 mg/kg/day and 50 mg/kg/day respectively).

Till now, we have been able to define some of the biochemical, morphological and behavioral changes associated with the toxicity of triethyltin which are independent of its oedematogenic effects on the cerebral tissue. It seems that the behavioral changes associated with the primary toxicity of triethyltin are due to the neurotoxic effects of triethyltin, and result as a consequence of the faulty conduction of the signals in the nerves fibers of the intoxicated animals. Thus the intoxicated animals do not react normally to electrical or other types of stimuli like heat and show reduced motility or increased tolerance to electroconvulsive treatments or heat induced algesia. It is important to mention, at this point, that all the primary or secondary toxicological manifestations described above for triethyltin are also observable in the animals after intoxicating them with another uncoupler of oxidative phosphorylation, hexachlorophene, and that the extract of Ginkgo biloba inhibits the neurotoxities of the two uncouplers in identifical manner. In addition, pentobarbital and other radical scavengers do not show any beneficial effects in these models in our experimental condition. The available information, till now, thus strongly suggest that the extract of Ginkgo biloba have some other biochemical properties which are not dependant on its radical scavenging property.

So far I have mentioned only those pharmacological properties of the extract with which our research group was directly involved. It can, however, not be overemphasized that the extract possesses other properties of interest for the therapy of cerebrovascular diseases. For example, the extract inhibits increased capillary permeability and membrane fragility, modulates functions of platelets, erythrocytes and other blood components, inhibits proteolytic enzymes and thromboxane synthetase and possesses many other pharmacological effects which will be discussed by other authors of this symposium.

CONCLUSION

In view of this broad spectrum of the pharmacological and biochemical activity profile of the extract it is essential to analyse the situation and to differentiate between those properties which are relevant for its therapeutic beneficial effects and those which are experimental artefacts. Such an analysis of the available data makes it evident that the broad activity profile of the extract is due to its effects on the three basic compartments of the body involved in the pathobiology of ischemic vascular diseases (see Fig.8). Our current understanding of the pathogenesis of such diseases clearly indicate that it is a dynamic and self-propagating process, which involves vasculature, blood and other bodyfluids as well as the metabolic and neurohormonal regulatory mechanisms of the injured tissue itself; which, in the case of cerebral diseases, is the brain tissue. Any disturbance in one of these systems triggers disturbances in the complex interplay between these systems and thus initiates a circulus viciosus. As a consequence, any rational therapy of such pathological conditions is only possible when individual components of the disease are properly taken care off. i.e. to say drugs which act only on a specific aspect of the ischemic disease can never be expected to give a global relief of the sufferings. In this respect the Extract of Ginkgo biloba by virtue of its multifactorial mode of action, offers an unique possibility of helping the patients with ischemic diseases. Results of the experiments conducted till now with different isolated constituents of the extract strongly indicate that this multifactorial mode of action of the drug is due to the fact that the extract represents a mixture of substances with different protective and curative properties against the various pathological processes involved in the progression of ischemic conditions.

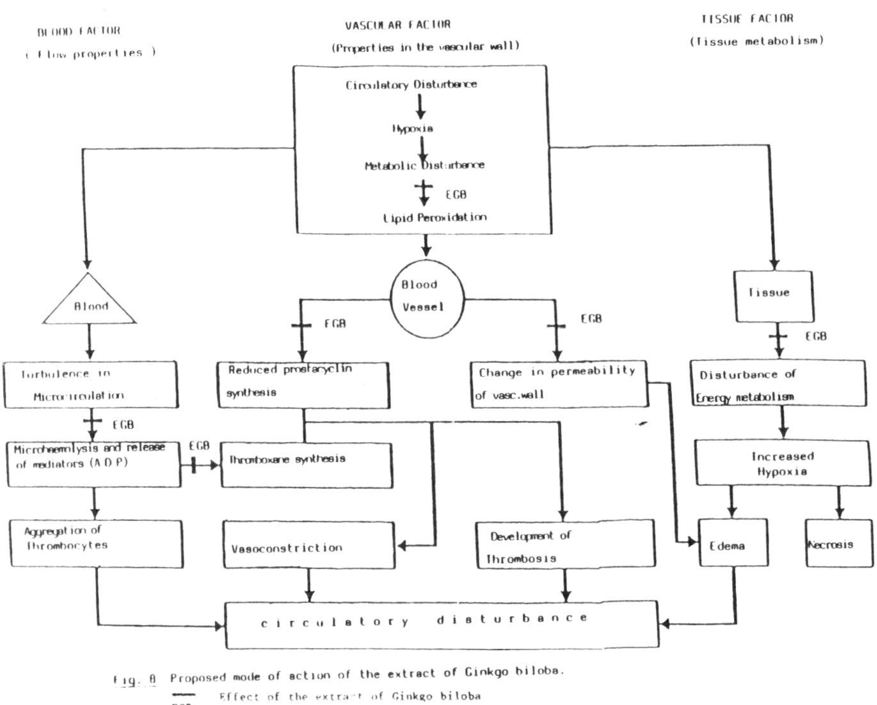

Fig. 8 : Proposed mode of action of the extract of Ginkgo biloba.
EGB Effect of the extract of Ginkgo biloba.

REFERENCES

1. Papers (1984) : presented in the "Clinical Studies" section of this symposium.
2. Chatterjee S.S. and Trunzler G. (1981) : Neue Èrgebnisse aus der Gingko-Forschung. Ärztezeitschrift Naturheilverf.22.
3. Peter H.J., Fisel J., Weisser W. (1966) : Zur Pharmakologie der Wirkstoffe aus Ginkgo biloba. Arzneim.Forsch.16.
4. Chatterjee S.S. and Gabard B. (1982) : Studies on the mechanism of action of an extract of Ginkgo biloba, a drug used for the treatment of ischemic Vascular Diseases.
Naunyn-Schmiedeberg's Arch. Pharmacol. 321, R 52.
5. De Feudis F.V. (1984) : This symposium.
6. Chatterjee S.S. and Gabard B. (1981) : Protective effect of an extract of Ginkgo biloba and other Hydroxyl Radical Scavengers against hypoxia.
Abstract: 8th Intern. Congress of Pharmacology (IUPHR).
Tokyo, July 19-24, page 483f (1981), Abstr. N°. 866.
7. Demopoulus H.B., Flamm E.S. and Seligman M.L. and al. (1977) : Antioxidant effects of barbiturates in Model. Membranes Underging Free Radical damage.
Acta Neurol. Scand. 56, 152-153.
8. De Feudis F.V. (1984) : This symposium.
9. Braquet P., Dily M., Bonhomme B. and Meyniel G. (1984) : Lipid peroxidation effects on isolated Rat. Retina.
Oxygen Radicals in Chemistry and biology.
Walter de Cruyter + Co. 913-918.
10. Braquet P., Dily M., Bonhomme B. and Meyniel G. (1982) : Peroxydation lipidique et activité électrique de la rétine isolée: effet protecteur de l'extrait de Ginkgo biloba.
Journées Internationales du Groupe Polyphénols - Toulouse 1-2 octobre.
11. Free Radicals in Medicine and Biology (1980) : Ed. D.H. Lewis and F. Del. Maestro.
Acta Physiolo Scandinavica, Supplementum 492.
12. Carpentier M.P. (1981) : Antioxidant effects on the prostaglandin endoperoxide synthelase product Profile.
Fed. Proc. 4o, 189-194.
13. Moncada S., Gryglewski R., Beenting S. and Vane J.R. (1976) : A lipid peroxide inhibits the enzyme in blood vessel microsomes that generates from prostaglandin endoperoxides the substance (prostaglandin x) which prevents platelet aggregation.
Prostaglandin 12, 715-737.
14. Herman A.G. and Beetens A.J. (1983) : Effects of vasoactive substances on prostacycline synthesis.
XV. Bad herrenalber angiologisches gespräch 22-23 April.
15. Karcher L., Zagermann P. and Krieglstein J. (1984) : Effect of an extract of Gingko biloba on rat brain energy metabolism in hypoxia.
Naunyn-Schmiedeberg's Arch. Pharmacol. 327, 31-35.
16. Poncin-Lafitte M.C., Rapin J. and Rapin J.R. (1980) : Effect of Ginkgo biloba on changes induced by quantitative cerebral microembolization in rats.
Arch. int. pharmacodyn. 243, 236-244.
17. Borzeix M.G. (1984) : This symposium.
18. Cahn J. (1984) : This symposium.

Effects of Ginkgo Biloba Extract on Organic Cerebral Impairment. A. Agnoli, J. R. Rapin, V. Scapagnini, W. V. Weitbrecht. John Libbey Eurotext Ltd

Some in vitro and in vivo actions of an extract of Ginkgo biloba (GBE 761)

F. V. DeFeudis, M. Auguet, S. Delaflotte, A. Hellegouarch, J. Baranes, M. Chapelat, M. Braquet, A. Etienne, K. Drieu, F. Clostre and P. Braquet

I.H.B.-IPSEN Institute for Therapeutic Research, 91940 Les Ulis, France

ABSTRACT

Studies were conducted in an effort to define some mechanisms which might underlie the therapeutic actions of Ginkgo biloba extract (GB 761). In vitro results indicated that GB 761 contracts rabbit aorta by releasing endogenous catecholamines and by potentiation of the contractile action of norepinephrine, and that GB 761 relaxes rabbit aorta by a mechanism which involves the release of endothelium-dependent relaxing factor(s). Treatments with GB 761 in vivo decreased both the acute and chronic phases of adriamycin-induced paw-inflammation, and partially reversed the increases in brain water, Na^+ and Ca^{2+} contents and the decrease in brain K^+ content that were associated with AA-induced cerebral infarct. These in vivo actions of GB 761 might be due to its free radical-scavenging action. Taken together, these results support the use of GB 761 in treating cerebral and peripheral vascular disorders and associated ischaemic conditions.

KEY WORDS

Ginkgo biloba extract (GB 761), rabbit aorta, endothelium-derived relaxing factor (EDRF), adriamycin, inflammation, arachidonate-induced cerebral infarct, norepinephrine, cardiovascular therapy, cerebral ischaemia, free radicals.

INTRODUCTION

Ginkgo biloba extract (GB 761), obtained from the dry green leaves of the tree Ginkgo biloba, is effective in treating various cardiovascular disorders (e.g., cerebrovascular insufficiency of the aged; cerebral ischaemia; peripheral vascular disease) (see e.g., Safi and Galley, 1977; Schwerdtfeger, 1981; Eckmann and Schlag, 1982; Pidoux et al, 1983). This extract contains glucosides (some of which are specific to GB 761), proanthocyanidines, terpenoid-derivatives (ginkgolides, bilobalide), and some common organic acids (e.g., vanillic acid, para-hydroxybenzoic acid). The diversity of substances present in GB 761 appears to be responsible for its polyvalent therapeutic action, which is essential for treating disorders of multifactorial origin.

Studies with experimental animals have added support for the therapeutic action of GB 761, in showing that this extract exerts beneficial effects in models of cerebral ischaemia, cerebral microembolization, and in different types of hypoxia (e.g., Larssen et al, 1978; Le Poncin Lafitte et al, 1980, 1982; Chatterjee and Gabard, 1981).

The experiments discussed herein were aimed at examining the effects of GB 761 on blood vessels, and its radical-scavenging action, both of which could contribute significantly to its therapeutic action.

MATERIALS AND METHODS

Studies on Rabbit Isolated Aorta

Male New Zealand rabbits, 1.8 - 2.6 kg, were used. In experiments on the contractile action of GB 761, some of the animals were pretreated with reserpine (5 mg/kg, i.m.) 18 hours before sacrifice.

Rabbits were killed by cervical dislocation, and their thoracic aortae were rapidly excised, placed in physiological medium (see below) at 37°C, and freed of superficial fat and connective tissue. Spiral strips of aorta (3 - 4 mm wide and 10 - 16 mm long) were prepared (Furchgott and Bhadrakom, 1953), and suspended vertically in 20-ml organ baths. Isometric contractions were recorded using force-displacement transducers (Narco Bio-Systems, F60) coupled to a polygraph (Narco Bio-Systems, MK IV or MK III). Strips were subjected to a tension of 2 g, and a one-hour equilibration period was allowed before commencing experiments.

The Krebs medium had the following composition (mM) : NaCl, 118; KCl, 4.7; $MgSO_4$, 1.17; KH_2PO_4, 1.18; $CaCl_2$, 2.5; $NaHCO_3$, 25; glucose, 11. It was continuously gassed with 95% O_2/5% CO_2; pH ≃ 7.4.

Cumulative concentration-response curves for GB 761, norepinephrine (NE) and tyramine were constructed (Van Rossum, 1963) using aortic strips that had been previously sensitized with NE (10^{-6} M). Antagonists or inhibitors of catecholamine uptake were added to the medium 5 minutes before the second series of cumulative concentrations: phentolamine (10^{-7} M) was tested against GB 761; cocaine (10^{-5} M) and desipramine (10^{-7} M) were tested against GB 761, NE and tyramine.

For studies of the possible potentiation of the contractile effects of NE, serotonin (5-HT) or dopamine (DA) by GB 761, the strips were sensitized by three successive exposures (at 31-min intervals) to agonists at concentrations which induced 80 - 90% of their maximal response (i.e., 10^{-6} M for NE and 5-HT; 3 x 10^{-5} M for DA). Two cumulative concentration-response curves were constructed. A 50-minute interval was allowed before constructing the second concentration-response curve. The action of GB 761 (100 μg/ml) was tested on the second curve, and it was given as a 5-minute pre-treatment.

The effect of GB 761 (0.3 - 6 mg/ml) of relaxing strips that had been pre-contracted with 60 mM KCl were also monitored.

For experiments on the possible involvement of endothelium-derived relaxing factor (EDRF) in GB 761-induced relaxation, the endothelium of some aortic strips was removed by mechanical means. Isometric contractions were recorded (see above); strips were pre-contracted with phenylephrine (PE, 10^{-7} M). The PD_2 value for carbachol and the PA_2 value for atropine were calculated (Van Rossum, 1963).

Studies with Adriamycin

The method of Siegel et al (1980), adapted to rats, was used. Groups of eight Sprague-Dawley rats (Charles River), 160 - 180 g, were used. At time zero, 0.25 mg/0.1 ml of an aqueous solution of adriamycin was injected into the left hind-paw of each animal. Inflammation of the paw was measured with a plethysmograph (Ugo Basile) before, and 5 days after, adriamycin injection. Percentage inflammation was calculated as follows :

$$\left(\frac{\text{volume at 1 hour (or on day 5) - initial volume}}{\text{initial volume}}\right)(100).$$

Drugs, dissolved in physiological saline, were administered orally or subcutaneously (5 ml/kg body weight) 2 hours before, and 24 hours and 48 hours after, adriamycin injection; i.e., 3 administrations. Sub-plantar injection of adriamycin produced a marked biphasic inflammatory response; the second phase was due to an increased vascular permeability caused by alterations of the membrane properties of endothelial cells (Siegel et al, 1980) that followed oxidative damage by the free radicals formed. As GB 761 exerts protective effects on membrane components (Etienne et al, 1980), its effect on the action of adriamycin was examined.

Arachidonate-Induced Cerebral Infarct

Forty male Sprague-Dawley rats, 220 - 280 g, were used. Test substances were administered 2 minutes before injection of sodium arachidonate (AA). Brain infarct was induced according to Furlow and Bass (1975). Under thiopental (35 mg/kg, i.v.) anaesthesia, a catheter was inserted into the left common carotid artery after ligation of the external branch. After recovery from anaesthesia, intra-carotid injection of AA (0.7 mg/rat) was performed at a rate of 0.3 ml/min; during this injection neurological symptoms (e.g., startles, lateral decubitus clonic-tonic discharges) were monitored, and finally the death-rate was recorded. Two hours after AA injection the rats were decapitated, their brains were rapidly removed and placed in a hot air bath for 3 days, and then the brain water content and cerebral Na^+, K^+ and Ca^{2+} levels were determined. Plasmatic malondialdehyde formation, an index of lipid peroxidation, was determined by measuring thiobarbituric acid-reactive (TBAR) material (Yagi, 1976).

Drugs and Reagents

Extract GB 761 (lot 127/128; Institut Henri Beaufour) was used. All other drugs and reagents were of the highest chemical purity.

RESULTS

Contractile Effects of GB 761 on Rabbit Isolated Aorta

GB 761 induced a reproducible, concentration-dependent contractile response in rabbit aortic strips ($EC_{50} \simeq 1.0 \pm 0.04$ mg/ml; n = 15) which was abolished by the α-adrenoceptor antagonist phentolamine (10^{-6} M); see also Auguet et al (1982b). Cocaine (10^{-5} M) and desipramine (10^{-7} M), blockers of neuronal catecholamine uptake, also inhibited the contractile effect of GB 761, but potentiated the contractile effect of NE and competitively inhibited that of tyramine (which acts by enhancing catecholamine release from nerve terminals); see also Auguet et al (1982b). The contractile effect of NE was slightly potentiated, but those of GB 761 and tyramine were markedly inhibited in aortic strips prepared from rabbits that had been pretreated with reserpine (which depletes catecholamine stores) (Fig. 1).

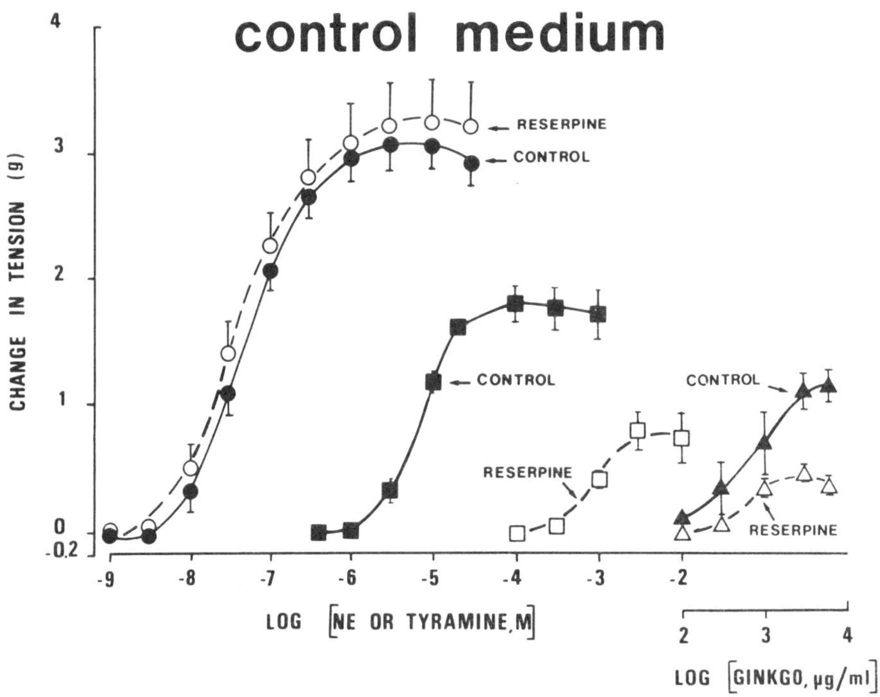

Fig. 1. Comparison of the contractile effects of GB 761, NE and tyramine on aortic strips prepared from control or reserpinized rabbits. Means ± S.E.M.; n = 4 for all points. Reproduced with permission from Auguet et al (1982b).

Results shown in Figure 2 indicate that a low concentration of GB 761 (100 μg/ml) potentiates the contractile effect of NE. The EC_{50} value for NE under control conditions was $7.4 \pm 0.4 \times 10^{-8}$ M; in the presence of GB 761, it was $3.6 \pm 0.2 \times 10^{-8}$ M (n = 5 in each case; $p < 0.001$; Student's *t*-test).

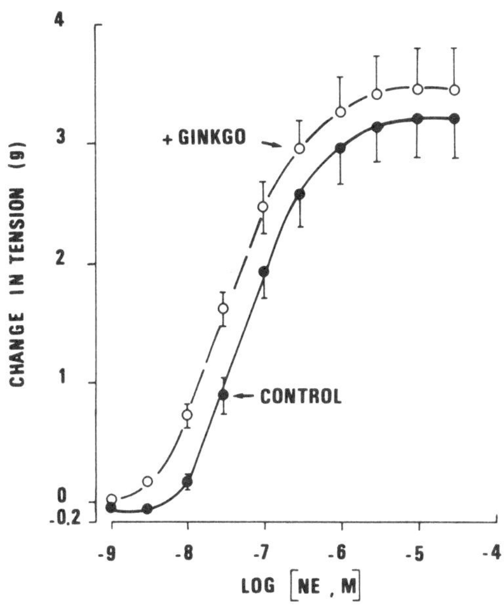

Fig. 2. Concentration-response curves for the contractile action of NE on rabbit aortic strips in the presence (○) or absence (●) of GB 761 (100 μg/ml), presented in terms of muscle tension. Means ± S.E.M.; 5 strips in each case. Reproduced with permission from Auguet et al (1982a).

Time-Dependent Relaxant Effect of GB 761 on Rabbit Aorta

In depolarizing (60 mM KCl) medium, which contracted aortic strips, GB 761 produced a concentration-dependent, time-dependent relaxation (Fig. 3). This effect of GB 761 differed markedly from its effect in normal medium (i.e., contraction; see above).

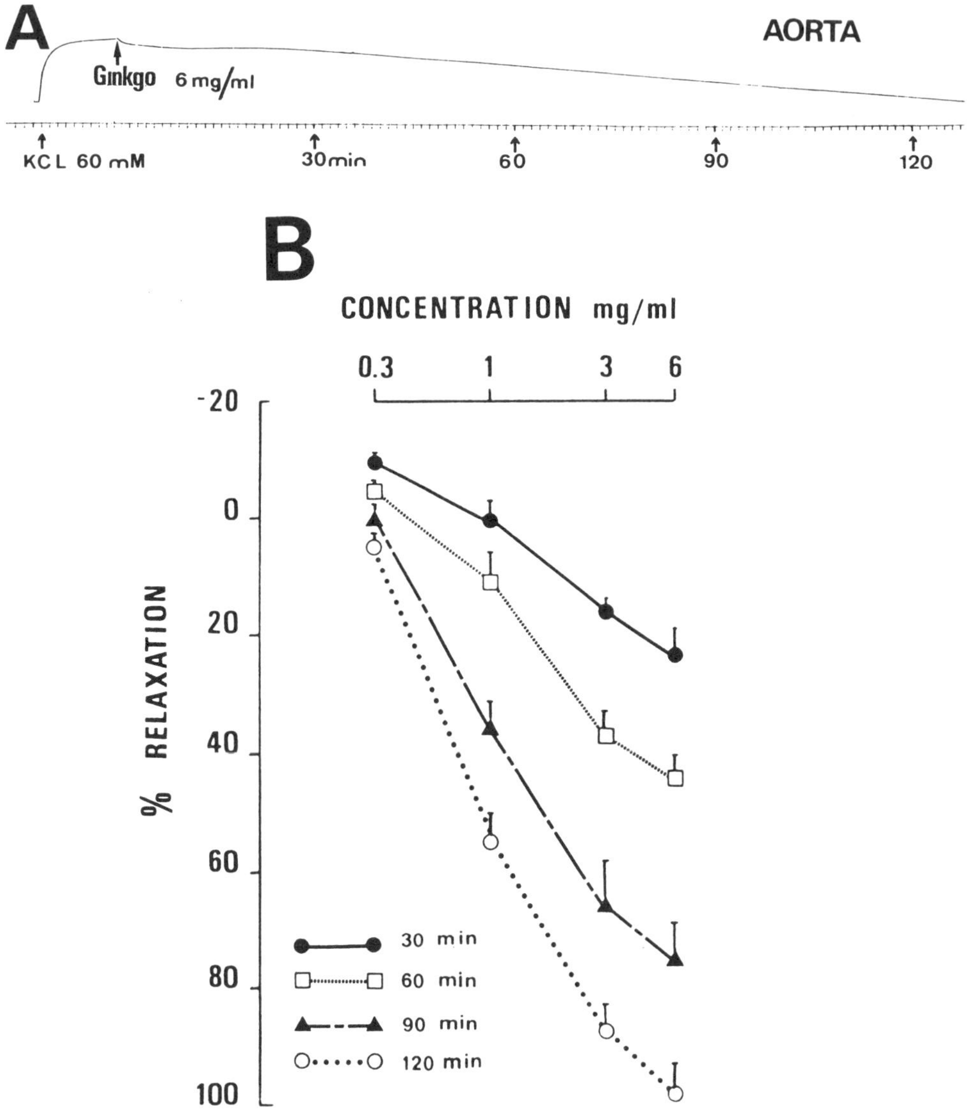

Fig. 3. Time-dependent relaxant action of GB 761 on isolated rabbit aortic strips that had been pre-contracted with KCl (60 mM). A, Single experiment with GB 761 (6 mg/ml); B, Concentration-response curves; means ± S.E.M.; 4 aortic strips. The effect of GB 761 had $IC_{50} \simeq 1.0$ mg/ml at 120 min. Reproduced with permission from Auguet et al (1984).

Dependence of GB 761-Induced Relaxation of Rabbit Aorta on EDRF

GB 761 (50 - 300 µg/ml), like carbachol ($PD_2 \simeq 6.2 \pm 0.1$), produced an endothelium-dependent relaxation of rabbit aortic strips (Fig. 4). The action of carbachol was competitively antagonized by atropine ($PA_2 \simeq 9.4 \pm 0.1$), but that of GB 761 was unaffected by atropine (10^{-7} M). Endothelium-dependent relaxations produced by both GB 761 and carbachol were blocked by 30-minute pretreatments with substances having lipoxygenase-inhibitor/anti-oxidant activity, such as nordihydroguaiaretic acid (NDGA), but were unaffected by agents that block cyclooxygenase activity (e.g., indomethacin, 10^{-5} M).

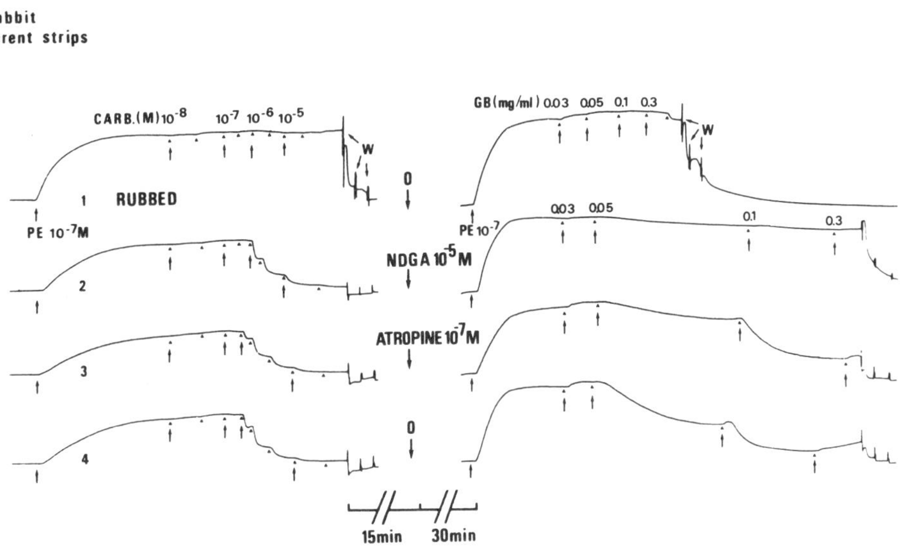

Fig. 4. Effects of de-endothelialization, NDGA and atropine on the relaxant action of GB 761 on rabbit isolated aorta. Left : Carbachol relaxed intact strips, but this action was abolished in the de-endothelialized (rubbed) preparation. Right: GB 761 relaxed intact strips. This action was not blocked by atropine, was partially blocked by NDGA, and was abolished by de-endothelialization. Other experiments showed that a 30-min pretreatment with a relatively high concentration of GB 761 (300 µg/ml) can inhibit endothelium-dependent relaxations produced both by carbachol and by GB 761, itself (Studies of S. Delaflotte et al, 1984; unpublished).

Studies with Adriamycin; Free-Radical Scavenging Action of GB 761

Diphenhydramine, cysteamine and GB 761 decreased the first phase of inflammation; diphenhydramine was more potent than cysteamine or GB 761 (Table 1). Concerning the second phase of inflammation (which involves membrane lipoperoxidation), N-acetylcysteine, cysteamine dephenhydramine, rutin and quercetin had no significant protective effect, whereas GB 761 was effective by both subcutaneous and oral routes (Table 1).

Table 1. Effects of various substances on adriamycin-induced inflammation in the rat

Lot of Animals	Dose (mg/kg)	Route of injection	% Inflammation 1 hour	% Inflammation 5 days
Control	-	-	66.0 ± 4.6	59.3 ± 7.6
N-Acetyl-L-cysteine	150	s.c.	66.7 ± 6.8	51.2 ± 4.6
GB 761	50	s.c.	68.6 ± 3.1	44.8 ± 6.6
GB 761	100	s.c.	73.4 ± 6.5	40.0 ± 4.9*
Control	-	-	32.0 ± 3.2	31.3 ± 4.1
Diphenhydramine	50[a]	s.c.	10.5 ± 1.4***	38.5 ± 3.3
Cysteamine	50	s.c.	22.1 ± 2.0*	28.8 ± 4.2
GB 761	50	s.c.	31.3 ± 4.1	19.1 ± 2.4*
Control	-	-	38.4 ± 3.1	33.2 ± 2.8
Rutin	200	s.c.	40.4 ± 2.0	35.2 ± 3.8
Quercetin	200	s.c.	31.3 ± 5.6	25.5 ± 5.0
GB 761	100	s.c.	34.5 ± 1.3	22.7 ± 2.7*
Control	-	-	37.3 ± 4.3	31.9 ± 3.6
GB 761	100	s.c.	34.9 ± 4.8	21.0 ± 4.2
GB 761	100	p.o.	32.5 ± 3.7	29.9 ± 3.3
GB 761	200	p.o.	25.3 ± 1.8*	21.8 ± 3.5*
GB 761	400	p.o.	27.1 ± 3.2*	20.5 ± 3.2*

Adriamycin (0.25 mg/0.1 ml) was injected into one hind-paw; 8 rats per lot. Test substances were given 2 h before, and 24 h and 48 h after adriamycin, except where otherwise noted. GB 761 = extract of *Ginkgo biloba* (lot 127/128).

* $p < 0.05$ and *** $p < 0.001$ by comparison with control animals (Student's *t*-test).

[a]Single dose, given 2 h before adriamycin.

Reproduced with permission from Etienne et al (1984).

Studies of AA-Induced Cerebral Infarct; Free Radical Scavenging Action of GB 761

Results shown in Table 2 revealed that AA-induced cerebral infarcts were associated with a pronounced loss of brain K^+, with increases in brain Na^+ and Ca^{2+}, and with cerebral oedema (increased % water). These changes were correlated with an increase in plasma malondialdehyde (TBAR-material). Surprisingly, SOD increased AA-induced damage. However, treatments with GSH-Px and GB 761 decreased cerebral oedema and led to partial recovery of cerebral Na^+ and K^+ levels. Only the highest dose of GB 761 affected TBAR-material (Table 2).

Table 2

Effects of free radical scavengers on brain electrolytes and malondialdehyde formation after injection of sodium arachidonate (AA)

Treatment	Dose (kg^{-1})	Brain water (%)	Brain electrolytes (m-mole/kg dry wt)			Plasma TBAR material (n-mole/ml)
			Na^+	K^+	Ca^{2+}	
Control	-	78.6 ± 0.1	237 ± 2	502 ± 4	4.2 ± 0.3	0.54 ± 0.06
AA Control	-	82.0 ± 0.4	381 ± 16	365 ± 12	6.1 ± 0.7	0.67 ± 0.12
SOD	0.5 mg	81.0 ± 0.4	354 ± 15	374 ± 8	5.6 ± 0.6	0.64 ± 0.11
SOD	1.0 mg	82.2 ± 0.5	394 ± 17	344 ± 16	5.8 ± 0.7	0.66 ± 0.12
SOD	2.0 mg	82.9 ± 0.5	412 ± 18	327 ± 15	6.3 ± 0.9	0.70 ± 0.11
GSH-Px	20,000 U	82.1 ± 0.4	357 ± 9	397 ± 13	6.0 ± 0.7	0.65 ± 0.10
GSH-Px	40,000 U	80.6 ± 0.4*	324 ± 10	444 ± 14	5.3 ± 0.4	0.60 ± 0.09
GSH-Px	80,000 U	80.2 ± 0.4*	299 ± 7*	461 ± 13*	5.1 ± 0.4*	0.59 ± 0.08
GB 761	5 mg	81.6 ± 0.5	344 ± 12	404 ± 14	5.9 ± 0.6	0.63 ± 0.09
GB 761	10 mg	80.5 ± 0.4*	305 ± 10	454 ± 14*	5.6 ± 0.6	0.59 ± 0.09
GB 761	20 mg	79.5 ± 0.4*	284 ± 8*	482 ± 14*	5.2 ± 0.5*	0.57 ± 0.07*

Means ± S.E.M.; 5 rats in all cases; *$p < 0.01$ between treated and AA control; all differences between control values and AA control values were significant ($p < 0.01$). SOD and GSH-Px were given i.v.; GB 761 was given i.p. two minutes before injection of AA.

Reproduced with permission from Etienne et al (1984).

DISCUSSION

Contractile and Relaxant Actions of GB 761 on Rabbit Aorta

Studies of the effects of drugs on the action of GB 761 on aortic strips of normal or reserpinized rabbits, and comparison of the contractile actions of GB 761 with those of NE and tyramine have revealed that contractions elicited by GB 761 involve, at least in part, a release of catecholamines from endogenous tissue stores. This was particularly evident from experiments which showed that inhibitors of catecholamine uptake (cocaine, desipramine) or reserpine pre-treatment of the animals inhibited contractions induced by GB 761 and tyramine while potentiating contractions induced by NE. A low molecular weight, "tyramine-like" substance (which might be present in GB 761) could account for this action.

Other experiments which showed that a low concentration of GB 761 (100 µg/ml), which did not evoke contraction by itself, could potentiate the contractile action of NE (but not those of 5-HT or DA) also indicated that GB 761 interacts with vascular catecholaminergic systems. This action of GB 761 might be due to an indirect mechanism: e.g., inhibition of catechol-O-methyltransferase or monoamine oxidase; inhibition of catecholamine uptake.

The findings that GB 761 elicited relaxations of aortic strips that had been pre-contracted either with high K^+ (Fig. 3) or with PE (Fig. 4), and especially that the relaxant effect of GB 761 is dependent upon an intact endothelium, are of major interest with respect to explaining the therapeutic actions of GB 761. These results indicate that GB 761 would relax arteries (blood vessels) which possess abnormally high tone thus supporting the use of GB 761 in treating certain vasospastic conditions, cerebral ischaemia and perhaps cerebrovascular insufficiency of the aged.

Mechanism of the Vasorelaxant Action of GB 761; Involvement of EDRF

Like carbachol, acetylcholine, substance P, ATP, thrombin, the antihypertensive agent hydrallazine, and many other substances (see e.g., Furchgott, 1983, 1984), GB 761 relaxes isolated rabbit aorta by an endothelium-dependent mechanism. Thus, GB 761, in causing relaxation, releases a substance(s) or activates a signal(s), tentatively defined as "endothelial-derived relaxing factor" (EDRF) which is responsible for relaxation of the underlying vascular smooth muscle cells. Studies with various enzyme inhibitors, anti-oxidants, free radical scavengers, and other substances have indicated that the "EDRF mechanism" is inhibited by substances possessing anti-oxidant activity and that EDRF is generated by a wide variety of substances possessing vasorelaxant activity (see e.g., Furchgott, 1983, 1984; Griffith et al, 1984). In this context, it seems noteworthy that GB 761 not only relaxes vasculature by a mechanism that involves EDRF, but that pre-treatment with higher concentrations of GB 761 can also block EDRF-related relaxation. Thus, GB 761 might interact with some endothelial membrane receptor to cause relaxation (one active principle) and might block EDRF action by its anti-oxidant, lipoxygenase-inhibitor, or free radical scavenging action (a separate active principle). A free radical could be involved in EDRF action, or EDRF could be a free radical (see e.g., Furchgott, 1984). In this regard, it is also noteworthy that the nitrovasodilators (whose relaxant action does not involve the endothelium) might act via the formation of the nitric oxide free radical, which would then activate guanylate cyclase, leading eventually to a dephosphorylation of myosin light-chain kinase and muscle relaxation (see e.g., Rapoport and Murad, 1983).

Studies of Adriamycin-Induced Inflammation and AA-Induced Cerebral Infarct

Free radical formation and the resulting lipid peroxidation appear to contribute significantly to the cell damage which occurs during cerebral hypoxia-ischaemia. Oxy-free radicals, such as superoxide radical ($O_2^{\cdot-}$) or hydroxyl radical ($OH^{\cdot}$) possess unusually high reactivity, and tend to attack neighboring molecules. Given these potentially devastating consequences of oxy-free radical formation, it is clear that all aerobic cells must possess an elaborate defense system. Such a defense system involves : superoxide dismutases which destroy $O_2^{\cdot-}$ and lead to the formation of O_2 and H_2O_2; catalase and glutathione peroxidase which destroy H_2O_2 and lipid peroxides, respectively; endogenous scavengers of free radicals (e.g., α-tocopherol flavonoids, ubiquinone, and thiol-containing substances; see Braquet et al, 1984).

The studies presented herein were aimed at comparing the effects of free radical scavengers (N-acetylcysteine, cysteamine, flavonoids, α-tocopherol, SOD, GSH-Px, and GB 761) using two *in vivo* models: adriamycin-induced inflammation; arachidonate-induced cerebral infarct. Results with the adriamycin test showed that GB 761, like cysteamine and the histamine-H_1 antagonist diphenhydramine, decreased the acute phase of inflammation. Regarding the second phase of adriamycin-induced inflammation, which involves the formation of free radicals (e.g., $O_2^{\cdot-}$) that alter membrane structure (Olsen et al, 1981), GB 761 was the only active substance; N-acetylcysteine and cysteamine were inactive.

With regard to AA-induced cerebral infarct, it is known that the massive platelet aggregation induced by intra-arterial injection of AA leads to multiple embolisms which occlude micro-circulatory endings. In human beings multiple cerebral infarct is generally associated with atherosclerotic lesions which involve platelet aggregation. This results in a well-defined neurological symptomatology and in marked perturbations of cerebral metabolism of both ischemic and oedematous origin, related to pathologic free radical reactions and lipoxygenase products. Hydroperoxides induced by the free radical actions on lipoxygenase pathways could selectively inhibit the formation of prostacyclin and result in the unopposed action of thromboxane. Experimental AA cerebral infarct, as studied, was accompanied by characteristic behavioral disturbances (e.g., jumping movements and clonic convulsions) and was followed in 80% of the cases by a pseudo-comatose state. Such disturbances were closely related to cerebral oedema; i.e., increases in brain water, Na^+ and Ca^{2+} contents and a decrease in brain K^+ content. Pre-treatment of the animals with GSH-Px or GB 761 reduced these biochemical changes. However, even at the highest doses used, recovery was only partial. It is proposed that a component(s) of GB 761 acted as a free radical scavenger, whereas GSH-Px destroyed the hydroperoxides that are produced during free radical-induced peroxidation.

REFERENCES

Auguet, M., DeFeudis, F.V. and Clostre, F. (1982a): Effects of Ginkgo biloba on arterial smooth muscle responses to vasoactive stimuli. Gen. Pharmac. 13, 169-171.

Auguet, M. DeFeudis, F.V., Clostre, F. and Deghenghi, R. (1982b): Effects of an extract of Ginkgo biloba on rabbit isolated aorta. Gen. Pharmac. 13, 225-230.

Auguet, M., Hellegouarch, A., Delaflotte, S., Baranes, J., DeFeudis, F.V., Clostre, F., Braquet, P. and Drieu, K. (1984): Effects of Ginkgo biloba extract on rabbit isolated blood vessels. In: Cerebral Ischaemia, eds. A. Bes, P. Braquet, R. Paoletti and B.K. Siesjö, pp. 347-354. Amsterdam, Elsevier.

Braquet, P., DeFeudis, F.V., Deby, C. and Braquet, M. (1984): Involvement of oxygen free radicals in biologic and pathologic processes. In: Cerebral Ischaemia, eds. A. Bes, P. Braquet, R. Paoletti and B.K. Siesjö, pp. 265-283. Amsterdam, Elsevier.

Chatterjee, S.S. and Gabard, B. (1981): Protective effect of an extract of Ginkgo biloba and other hydroxyl radical scavengers against hypoxia. In: Proc. VIII Int. Congr. Pharmacol., Tokyo, Abstr. 866.

Eckmann, F. and Schlag, H. (1982): Etude contrôlée, à double insu, de l'activité de l'extrait de Ginkgo biloba chez des malades atteints d'insuffisance cérébrale chronique. Fortschr. Med. 31-32, 1474-1478.

Etienne, A., Baranes, J., Hecquet, F., Hellegouarch, A. and Clostre, F. (1980): Effet stabilisateur de membrane d'un extrait de Ginkgo biloba. Planta Medica 237, 6-11.

Etienne, A., Chapelat, M., Braquet, M., Clostre, F., Drieu, K., DeFeudis, F.V. and Braquet, P. (1984): In vivo studies of free radical scavenging activity; relation to cerebral ischaemia. In: Cerebral Ischaemia, eds. A. Bes, P. Braquet, R. Paoletti and B.K. Siesjö, pp. 379-384. Amsterdam, Elsevier.

Furchgott, R.F. (1983): Role of endothelium in responses of vascular smooth muscle. Circ. Res. 53, 557-573.

Furchgott, R.R. (1984): The role of endothelium in the responses of vascular smooth muscle to drugs. Annu. Rev. Pharmac. Toxicol. 24, 175-197.

Furchgott, R.F. and Bhadrakom, S. (1953): Reactions of strips of rabbit aorta to epinephrine, isopropylarterenol, sodium nitrite and other drugs. J. Pharmac. Exp. Ther. 108, 129-143.

Furlow, T.W. and Bass, N.H. (1975): Science 187, 658-661.

Griffith, T.M., Edwards, D.H., Lewis, M.J., Newby, A.C. and Henderson, A.H. (1984): The nature of endothelium-derived vascular relaxant factor. Nature (Lond.) 308, 645-647.

Larssen, R.G., Dupeyron, J.P. and Boulu, R.G. (1978): Modèle d'ischémie cérébrale expérimentale par microsphères chez le rat. Etude de l'effet de deux extraits de Ginkgo biloba et du naftidofuryl. Thérapie 33, 651-660.

Le Poncin Lafitte, M., Martin, P., Lespinasse, P. and Rapin, J.R. (1982): Ischémie cérébrale après ligature non simultanée des artères carotides chez le rat: Effet de l'extrait de Ginkgo biloba. Sem. Hôp. Paris, 58, 403-406.

Le Poncin Lafitte, M., Rapin, J. and Rapin, J.R. (1980): Effects of Ginkgo biloba on changes induced by quantitative cerebral microembolization in rats. Arch. Int. Pharmacodyn. 243, 236-244.

Olsen, R.D., Mac Donald, J.S., Van Boxtel, C.J., Boerth, R.C., Harbison, R.D., Slomin, A.E., Framan, R.W. and Oates, J.A. (1981): Mechanism of adriamycin cardiotoxicity; evidence for oxidative stress. Life Sci. 29, 1393-1401.

Pidoux, B., Bastien, Cl. and Niddam, S. (1983): Etude à double insu, face au placebo, des effets de l'extrait de Ginkgo biloba sur l'état clinique et l'EEG quantifié. J. Cereb. Blood Flow Metab. 3 (Suppl. 1) 5556-5557.

Rapoport, R.M. and Murad, F. (1983): Endothelium-dependent and nitro-vasodilator-induced relaxation of vascular smooth muscle: Role of cyclic GMP. J. Cyclic Nucleotides & Prot. Phosph. Res. 9, 281-296.

Safi, N. and Galley, P. (1977): Tanakan et cerveau sénile. Etude radiocirculographique. Bordeaux Med. 1, 171-176.

Schwerdtfeger, F. (1981): Traitement des vertiges par l'extrait de Ginkgo biloba, contrôles électronystagmographiques. Therapiewoche 31, 8658-8667.

Siegel, D.M., Giri, S.N., Scheinholtz, R.M. and Schwartz, L.W. (1980): Characteristics and effect of anti-inflammatory drugs on adriamycin induced inflammation in the mouse paw. Inflammation,4 233-248.

Van Rossum, J.M. (1963): Cumulative dose-response curves. II. Technique for the making of dose-response curves in isolated organs and the evaluation of drug parameters. Arch. Int. Pharmacodyn. 143, 299-330.

Yagi, K. (1976): Biochem. Med. 15, 212-213.

Effects of Ginkgo Biloba Extract on Organic Cerebral Impairment. A. Agnoli, J. R. Rapin, V. Scapagnini, W. V. Weitbrecht. John Libbey Eurotext Ltd © 1985

The effects of chronic, oral Ginkgo biloba extract administration on neurotransmitter receptor binding in young and aged Fisher 344 rats

J. E. Taylor

Biomeasure Inc. Hopkinton, MA, U.S.A.

ABSTRACT

The effect of Ginkgo biloba extract (GBE) on neurotransmitter function is studied in several brain regions as a possible explanation of his action on cerebral aging deficits. GBE is given orally at a dose of 100 mg/kg/day for 28 days to male Fisher 344 rats divided in two groups aged of 3 or 24 months. After sacrifice, membrane fractions of various brain regions were prepared. Neurotransmitters receptor binding were determined by incubating preparation with various concentration of the radio labeled receptor ligands. Analyse is based on linear regression of Scatchard. Results show that chronic treatment with GBE increases binding and density of the muscarinic acetylcholine receptor in the hippocampus of the aged rat. This action may contribute to some of the therapeutic actions of GBE especially on the cognitive and memory deficits associated with aging.

KEYWORDS

Ginkgo biloba extract - orally chronic treatment - ex vivo binding study - brain muscarinic receptors - acetylcholine.

INTRODUCTION

A number of studies have reported changes in neurotransmitter function in the central nervous system which appear to be associated with the aging process (McGeer, 1981 ; Roth and Hess, 1982). Some of the most prominent and reproducible changes have been shown to occur with respect to muscarinic cholinergic systems. For example, decreased choline acetyltransferase (CAT) activity has been observed in several brain regions of patients suffering from Alzheimer's dementia (Perry and al., 1977 ; Davis and Yamamura, 1978). Reduced muscarinic receptor binding has also been reported in the hippocampus of Alzheimer's patients and in the brains of aging rodents (Reisine and al., 1980 ; Norberg and Winblad, 1981 ; Nordberg and Wahlstrom 1981 ; Morin and Westerlain, 1980). In addition, changes in adrenergic and opiate receptor binding have been reported in the literature (Roth and Hess, 1982).

Our interest in the behavioral and neurochemical actions of the Ginkgo biloba extract (GBE) and the aging process led us to examine the effects of chronic GBE treatment on the in vitro binding characteristics of several neurotransmitter receptors in the brains of young and old rats.The results show that

GBE elicited an increase in muscarinic acetylcholine receptor binding in the hippocampus of the aged rat. Drug-induced changes in receptor binding for the other neurotransmitter receptors was not observed.

METHODS

Male Fisher 344 rats (3 or 24 months old) were administered GBE in their drinking water at a concentration calculated to provide a dose of 100 mg/kg/day for 28 days. Twenty four hours after the last dose the rats were sacrificed, and crude membrane fractions of various brain regions were prepared. Neurotransmitter receptor binding was determined by incubating aliquots of the membrane preparation with various concentrations of the radio-labeled receptor ligands. Assays were terminated by rapid filtration through Whatman GF/B filters under reduced pressure, and the bound radioactivity trapped on the filters was counted by liquid scintillation spectrometry. Linear regression of Scatchard plots were used to determine the equilibrium dissociation constants (kd), and the maximum binding capacities (Bmax). Statistical analysis of the data was done using the Student's t-test.

RESULTS

The effects of chronic GBE administration on (^{3}H) QNB (quinuclidinyl benzylate binding to hippocampal muscarinic receptors are shown in Table I.

Table I. The effect of Chronic GBE Administration on the Specific Binding of (^{3}H) QNB to Hippocampal Muscarinic Receptors

Animal	(n)	Bmax (fmol/mg tissue $\pm$ S.E.M.)
control-3 month	(8)	133.7 $\pm$ 4.2
control-24 month	(5)	108.8 $\pm$ 6.8*
+GBE-3 month	(6)	140.9 $\pm$ 3.5
+GBE-24 month	(6)	127.6 $\pm$ 3.7

*Significantly different from control - 3 month : GBE-3 month and GBE-24 month, $p < 0.05$

With regard to the control group, there was a significant decrease (-20%) in the Bmax of the aged animals compared to the young. The results confirm data previously reported (Nordberg and Winblad, 1981 ; Nordberg and Wahlstrom, 1981 ; James and Kanungo, 1976 ; Morin and Westerlain, 1980) that the density of muscarinic acetylcholine receptors in various brain regions decreases during aging.

A small increase in the number of muscarinic receptors was observed at both ages following chronic GBE treatment. This increase was most obvious in the aged animals and was statistically-significant. The Bmax value for the GBE-24 month group was similar to that observed for the control-3 month group suggesting that chronic GBE administration to the aged animal normalizes the density of muscarinic receptors.

The effects on other brain receptors are shown in Table II. Although there were some age-dependent changes (i.e. beta-adrenergic), chronic GBE administration had no effect on these neurotransmitter receptors.

Table II. The Effect of Chronic GBE Administration on Adrenergic, Dopamine, Serotonin, and Opiate Receptors

Receptor (Brain Region)	Bmax (fmol/mg tissue ± S.E.M.) Control (3 month)	Control (24 month)	+GBE (3 month)	+GBE (24 month)
$alpha_1$-adrenergic (c. cortex)	8.9 ± 0.3	9.0 ± 0.6	9.7 ± 0.5	8.4 ± 0.9
$alpha_2$-adrenergic (c. cortex)	5.6 ± 0.3	4.9 ± 0.6	6.0 ± 0.1	5.3 ± 0.40
beta-adrenergic (c. cortex)	8.1 ± 0.6	6.7 ± 0.3*	9.3 ± 0.5	6.7 ± 0.9*
$dopamine_2$ (striatum)	16.1 ± 1.9	13.7 ± 1.0	12.0 ± 0.7	15.1 ± 1.7
5-HT_2 (c. cortex)	12.1 ± 1.2	11.0 ± 1.2	12.7 ± 1.1	11.5 ± 1.3
opiate, mu (midbrain)	6.7 ± 0.5	6.7 ± 0.06	8.2 ± 0.9	10.9 ± 1.1

* Significantly different than control - 3 month, $p < 0.05$

CONCLUSION

Extensive evidence from in vivo and in vitro experiments has supported the idea that muscarinic cholinergic underactivity may account for some of the cognitive and memory deficits associated with aging (Davis and Yamamura, 1978 ; Mc Geer, 1981). The present data confirm the results of others that postsynaptic receptor changes may contribute to the decline in central cholinergic function with aging. We have also shown that chronic treatment with an extract of Ginkgo biloba increases the apparent muscarinic receptor density in the hippocampus of the aged Fisher 344 rat. This action may contribute to some of the therapeutic actions of Ginkgo.

REFERENCES

1. Davis, K.L. and Yamamura, H.I. (1978) : Cholinergic underactivity in human memory disorders. Life Sci. 23, 1729-1734.
2. James, T.C. and Kanungo, M.S. (1976) : Alterations in atropine sites of the brain of rats as a function of age. Biochem. Biophys. Res. Commun. 72, 170-175.
3. McGeer, E.G. (1981) : Neurotransmitter systems in aging and senile dementia. Prog. Neuro-Psychopharmacol. 5, 435-445
4. Morin, A.M. and Westerlain, C.G. (1980) : Aging and rat brain muscarinic receptors as measured by quinuclidinyl benzilate binding. Neurochem. Res. 5, 301-308.
5. Nordberg, A. and Wahlstrom, G. (1981) : Cholinergic receptors and aging. Abstr. Eight. Int. Congr. Pharmacol. 827.
6. Nordberg, A. and Winblad B. (1981) : Cholinergic receptors in human hippocampus regional distribution and variance with age. Life Sci. 29, 1937-1944.
7. Perry, T.L., Gibson, P.H., Blessid, G., Perry, R.H. and Tomlinson, B.E. (1977): Neurotransmitter abnormalities in senile dementia. J. Neurol. Sci. 34, 247-265.
8. Reisine, T.D., Pedigo, N.W., Meiners, C., Igbal, K. and Yamamura, H.I. (1980) : Alzheimer's disease : studies on neurochemical alterations in the brain : Aging of the Brain and Dementia (Amaducc, L., Davison, A.N. and Antuono P. eds). Raven Press, New-York, 147-150.

9. Roth, G.S. and Hess, G.D. (1982) : Changes in the mechanisms of hormone and neurotransmitter action during aging : Current status of the role of receptor and post-receptor alterations. A review, Mech. of ageing and Develop. 20, 175-194.

Effects of Ginkgo Biloba Extract on Organic Cerebral Impairment. A. Agnoli, J. R. Rapin, V. Scapagnini, W. V. Weitbrecht. John Libbey Eurotext Ltd © 1985

Behavioral effects of Ginkgo biloba extract

G. Continella and F. Drago

Institute of Pharmacology, Surgery and Medicine School, University of Catania

The extract of Gingko Biloba is known to exert a number of actions at the level of the cardiovascular system, mainly rapresented by vasodilation. This is accompained by some other peripheral effects, such as the slowing down of the intestinal transit and an increase of diuresis. Due to the activity of Gingko Biloba extract on the smooth muscles, hypotensive effects have also been observed.

At the level of the central nervous system, Gingko Biloba extract shows a favourable action on the electrical activity of cortical tissue suffering by experimental oedema.

The present experiments were aimed at investigating by behavioral techniques the central effects of Gingko Biloba extract in adult and old male rats. All the experiments have been performed on male albino rats, aging 8 and 24 months, in acute (intraperitoneal injection 1 hr before the test) and subchronic (8 days subcutaneous treatment before the testing day) experimental conditions. Gingko Biloba extract was administered at two dose levels: 25 and 50 mg/Kg.

The present investigations include:

- the study of the effects of Gingko Biloba extract on the acquisition of an active avoidance conditioned behavior, performed by the shuttle-box (single session) technique;
- the study of the effects of Gingko Biloba extract on the retention of a passive avoidance behavior, studied by the dark box technique;

- the study of the effects of Gingko Biloba extract on the immobility score in the despair test.

The results of the above experiments may be summarized as follows:

- acquisition of the active avoidance behavior: after acute administration of Gingko Boloba extract, no difference was found in the behavior of adult rats, while the acquisition ability of old animals -with the higher dose- was slightly but significantly improved (Fig.1). After a subchronic treatment, adult rats showed no significant behavioral modifications, while old animals, injected with both doses of Gingko Biloba extract, exhibited higher acquisition levels than their controls (Fig.2).
- retention of the passive avoidance behavior: after acute injection of Gingko Biloba extract, adult rats did not show any behavioral difference from their controls (Fig.3). Old rats injected with 50 mg/Kg of Gingko Biloba extract exhibited a better performance than their controls in the 24 hr passive avoidance retention test. Such an improvement was not found in the 48 hr passive avoidance retention test. After a subchronic treatment, old rats injected with both 25 and 50 mg/Kg Gingko Biloba extract showed an improved performance at the 24 hr test. At the 48 hr test, the old rats injected with the higher dose continued to show an improved behavior (Fig.4).
- despair test: nodifference was found in the immobility of adult and old rats when injected acutely with Gingko Biloba extract (Fig.5). After subchronic treatment, old rats administered with 50 mg/Kg Gingko Biloba extract showed a shortened immobility score (Fig.6).

Fig.1

EFFECTS OF THE ACUTE INJECTION OF GINGKO BILOBA EXTRACT ON THE ACQUISITION OF THE SHUTTLE-BOX ACTIVE AVOIDANCE BEHAVIOR OF ADULT AND OLD RATS

Group	Animals	(n)	Treatment
1	Young rats	(10)	Vehicle i.p.
2	Old rats	(10)	Vehicle i.p.
3	Young rats	(10)	GB 25 mg/Kg i.p.
4	Young rats	(10)	GB 50 mg/Kg i.p.
5	Old rats	(10)	GB 25 mg/Kg i.p.
6	Old rats	(10)	GB 50 mg/Kg i.p.

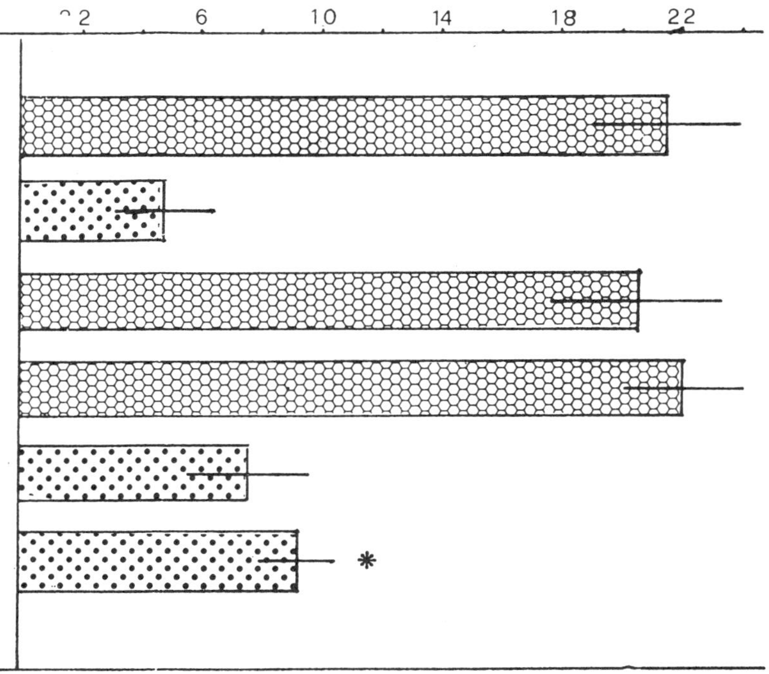

* $p < 0.05$ compared to group 2 (Student's t-test)

Fig. 2

EFFECTS OF THE SUBCHRONIC (8 DAYS) ADMINISTRATION OF GINGKO BILOBA EXTRACT ON THE ACQUISITION OF THE SHUTTLE-BOX OF ADULT AND OLD RATS

Group	Animals	(n)	Treatment
1	Young rats	(10)	Vehicle s.c.
2	Old rats	(10)	Vehicle s.c.
3	Young rats	(10)	GB 25 mg/Kg/die
4	Young rats	(10)	GB 50 mg/Kg/die
5	Old rats	(10)	GB 25 mg/Kg/die
6	Old rats	(10)	GB 50 mg/Kg/die

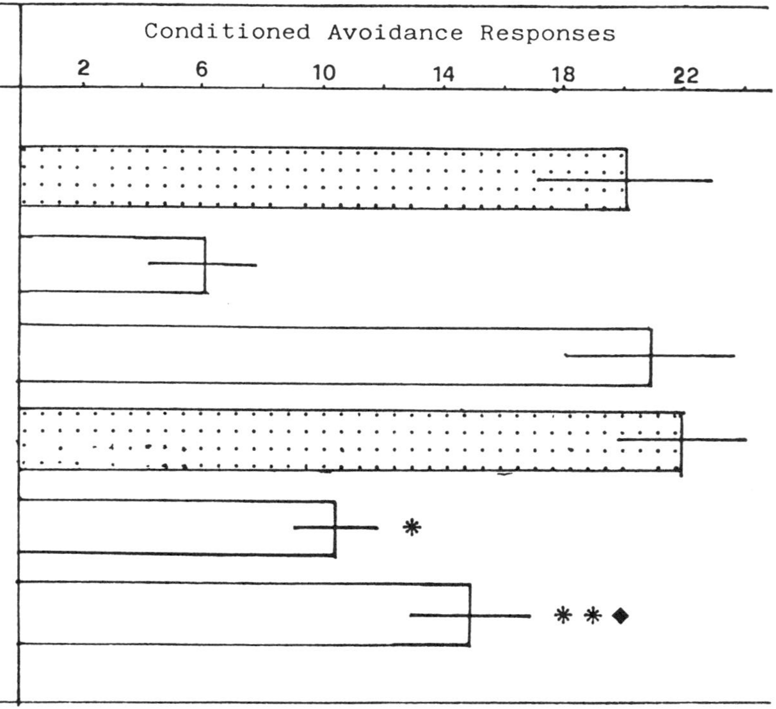

* $p < 0.05$ compared to group 2 (Student's t-test)

** $p < 0.01$ compared to group 2 (Student's t-test)

◆ $p < 0.05$ compared to group 5 (Student's t-test)

Fig. 3

EFFECTS OF THE ACUTE INJECTION OF GINGKO BILOBA EXTRACT ON THE RETENTION OF PASSIVE AVOIDANCE BEHAVIOR OF ADULT AND OLD RATS

Animals	(n)	Treatment	Avoidance latency (min)	
			24 hr	48 hr
Young rats	(10)	Vehicle i.p.	85.2 ± 6.2	80.3 ± 5.3
Young rats	(10)	GB 25 mg/Kg i.p.	83.3 ± 8.8	81.2 ± 5.0
Young rats	(10)	GB 50 mg/Kg i.p.	91.8 ± 8.1	85.0 ± 7.1
Old rats	(10)	Vehicle i.p.	5.0 ± 2.0	6.0 ± 2.1
Old rats	(10)	GB 25 mg/Kg i.p.	9.3 ± 1.2	9.5 ± 3.0
Old rats	(10)	GB 50 mg/Kg i.p.	15.4 ± 3.0*	10.4 ± 3.1

Values are means of avoidance latencies in sec ± SE.

* $p < 0.05$ compared to the control old rats (Student's t-test)

Fig.4

EFFECTS OF THE SUBCHRONIC ADMINISTRATION (8 DAYS) OF GINGKO BILOBA EXTRACT ON THE RETENTION OF PASSIVE AVOIDANCE BEHAVIOR OF ADULT AND OLD RATS

Animals	(n)	Treatment	Avoidance latency (min)	
			24 hr	48 hr
Young rats	(10)	Vehicle s.c.	83.3 ± 9.1	90.1 ± 5.4
Young rats	(10)	GB (25 mg/Kg/die s.c.)	85.8 ± 5.4	76.4 ± 4.1
Young rats	(10)	GB (50 mg/Kg/die s.c.)	79.0 ± 8.8	80.3 + 9.5
Old rats	(10)	Vehicle s.c.	7.1 ± 2.5	5.4 + 3.0
Old rats	(10)	GB (25 mg/Kg/die s.c.)	15.5 ± 2.1*	9.8 ± 3.1
Old rats	(10)	GB (50 mg/Kg/die s.c.)	25.1 ± 3.5**°	19.5 + 4.1*

Values are means of avoidance latencies in sec ± SE.

* $p < 0.05$ compared to the control old rats (Student's t-test)

** $p < 0.01$ compared to the control old rats (Student's t-test)

° $p < 0.05$ compared to the 25 mg/Kg treated old rats (Student's t-test)

Fig. 5

EFFECTS OF THE ACUTE INJECTION OF GINGKO BILOBA EXTRACT ON THE IMMOBILITY OF ADULT AND OLD RATS IN THE DESPAIR TEST

Animals	(n)	Treatment	Immobility (in sec)
Young rats	(8)	Vehicle i.p.	200.5 ± 10.1
Old rats	(8)	Vehicle i.p.	250.8 ± 15.2
Young rats	(8)	GB (25 mg/Kg i.p.)	188.0 ± 15.8
Young rats	(8)	GB (50 mg/Kg i.p.)	190.5 ± 13.4
Old rats	(8)	GB (25 mg/Kg i.p.)	230.0 ± 8.7
Old rats	(8)	GB (50 mg/Kg i.p.)	215.3 ± 16.1

Values are means ± S.E.

Fig. 6

EFFECTS OF SUBCHRONIC ADMINISTRATION OF GINGKO BILOBA EXTRACT ON THE IMMOBILITY OF YOUNG AND OLD RATS IN THE DESPAIR TEST

Animals	(n)	Treatment	Immobility (in sec)
Young rats	(8)	Vehicle s.c.	208.0 ± 8.1
Old rats	(8)	Vehicle s.c.	255.3 ± 16.0
Young rats	(8)	GB (25 mg/Kg/die s.c.)	195.3 ± 10.6
Young rats	(8)	GB (50 mg/Kg/die s.c.)	189.5 ± 15.4
Old rats	(8)	GB (25 mg/Kg/die s.c.)	218.5 ± 12.6
Old rats	(8)	GB (50 mg/Kg/die s.c.)	208.4 ± 11.3 *

Values are means ± SE

* $p < 0.05$ compared to the control old rats (Student' t-test)

Effects of Ginkgo Biloba Extract on Organic Cerebral Impairment. A. Agnoli, J. R. Rapin, V. Scapagnini, W. V. Weitbrecht. John Libbey Eurotext Ltd © 1985

Effects of Ginkgo biloba extract (GBE) on the acute phase of cerebral ischemia due to embolisms

J. Cahn

Département de Thérapeutique Expérimentale, SIR International, 6 rue Blanche, 92120 Montrouge, France

SUMMARY

The injection into a common carotid artery of fragments of an autologous clot results, in the dog, in a multiple cerebral infarction which progresses in 3 phases ; the first one which lasts approximately 15 minutes corresponds to the hyperhemix period ; the second one lasts from 30 to 120 min and corresponds to the progressive constitution of the post-embolic disease ; the last one corresponds to an hemodynamic and metabolic failure ; it is characterized by a deep hypoperfusion concomitant with a marked reduction of the oxygen availability, the $cvpO_2$ values reaching the lethal threshold according to Grote et al (1971). The I.V. perfusion of GBE from the begining of the third period prevents the evolution of this syndrome, in particular, CBF is improved as well as the oxygen supply.

KEY WORDS

Dog, multiple cerebral infarction, blood clot embolisms, CBF, oxygen availability, Ginkgo biloba extract.

INTRODUCTION

The common feature originating a multiple cerebral infarct in human beings is the splitting of an atheromatous plaque whose fragments occlude small cerebral arteries.

The injection into a common carotid artery of small embolisms obtained from a clot produced with autologous blood, mimics this clinical situation.

Area of hyperhemia may exist in the acute stage of stroke as first mentionned by N.A. Lassen (1966) as the "luxury perfusion syndrome". In fact, focal cerebral hyperhemia observed in human beings seems to be divided into 3 groups : borderzone hyperhemia, post-ischemic hyperhemia and remote hyperhemia (T.S. Olsen et al 1981). Ischemic lesions are responsible for the neurologic deficit observed in patients with stroke but in fact, cerebral infarcts of more than 24 hours are made up of 2 essentially different parts : one is ischemic, the other hyperhemic.

The experimental model resulting from the intra-carotid injection of blood embolisms corresponds to the clinical situation in that it associates both reactive hyperhemia and hypoperfusion-induced ischemia.

MATERIAL AND METHODS

Fourteen mongrel dogs of both sex, weighing about 30 kg were anesthetized with I.V. α-chloralose. Arterial blod was drawn from each dog to obtain the clot which was then mashed and suspended in the autologous serum ; the further injection of these fragments was performed under a volume àf 5 ml through a catheter inserted into a common carotid artery. Dogs were intubated by means of a tracheal tube coated with Tronothane R and when the reanimation started they were ventilated by means of a Bird Mark VIII respirator with a mixture of air + O_2. The end-tidal CO_2 was continuously controlled with a CO_2 test (Jaeger) connected with a catheter inserted into the endotracheal tube. Electromagnetic flowmeters (Statham SP 7517) were inserted into both left and right maxillar veins, after ligation of their facial afferences, in order to continuously measure the venous CBF. The cerebral venous pressure was measured by means of a catheter inserted into the left auricular vein and connected with an electromanometric cell (Statham P 23 BB - 0-6.4 kPa). MABP as well as systolic and diastolic pressure were monitored from a catheter inserted into a femoral artery by means of an electromanometric cell (Statham GB 6.5-30 kPa). These manometric cells were connected with pre-amplifiers and blood pressure computers (Gould). The EKG was continuously recorded in L II as well as the body temperature ("TC 100 Electrotherm" digital thermometer). All these parameters were continuously recorded on a Brush Mark 200 recorder. A needle catheter (Becton Dickinson) was inserted into the right maxillar vein as well as into the right carotid artery. Blood gases were measured in order to maintain $paCO_2$ ranged between 4.65 and 5.32 kPa. The needle catheters inserted into the common carotid artery and the upper right cerebral vein, were connected with two autoanalyzers Technicon, including two proportioning pumps, two dialyzers systems, two fluorimeters, two photometers and two recorders. Arterial and venous lactate determination were realized after dialysis by fluorimetric technics according to Y. Minaire, F. Foucheraud and C. Studievic's enzymatic method (Technicon Symposium 1965) while glucose determinations were performed by photocolorimetry according to P. Trinder's GOD/PAP enzymatic method.

Ginkgo biloba extract (IPS 200) was perfused I.V. (1 $ml.min^{-1}$) at the dose of 1 $mg.kg^{-1}$ for 20 min followed by 0.5 $mg.kg^{-1}$ every 60 minutes until the end of the experiment. Control dogs were given the same volume of solvent (saline).

RESULTS

The multiple cerebral infarction progresses in 3 different steps.

The first one develops within the 15 minutes following the injection of the clot. Outside of any change in MABP, there takes place a large and significant increase in CBF, the maximum being at 8 minutes post-injection. Over this transient period, the brain compensates this tremendous hyperhemia since it reduces the glucose extraction rate (glucER) and in contrast the arterio-venous difference for lactate increases showing thus that hyperhemia does not result from increased metabolic demands (table 1).

Table 1 : Cerebral hemodynamics and metabolism over the hyperhemic-period

Time (min) post injection	N	MABP (kPa)	CBF (ml.min-1 100 g-1)	GlucER (%)	(A-V) Gluc (mmol.1-1)	(A-V) Lact (mmol.1-1)
0	14	18.4 ± 0.43	192 ± 26.2	8.4 ± 1.12	0.35 ± 0.042	- 0.001 ± 0.057
8	14	18.0 ± 0.39	337 ± 39.4 (*)	6.4 ± 0.78 (*)	0.27 ± 0.033 (*)	- 0.09 ± 0.043 (*)
15	14	18.1 ± 0.33	236 ± 32.1	6.7 ± 1.15 (*)	0.28 ± 0.052	- 0.06 ± 0.070

(*) $P \leqslant 0.05$ according to Student's "t" test or to Cochran's "t" test with regards to time 0.

The second steps corresponds to the progressive constitution of the "disease" and lasts approximately from 30 to 120 min post-injection, according to each subject. This phase is mainly characterized by a gradual reduction of CBF which declines by 72 percent on an average, whereas the cerebral vascular resistance (CVR) increases by more than 200 percent while the cerebral perfusion pressure only drops by 18 percent ; this is thus a low CBF syndrome with a subnormal perfusion pressure. On a biochemical point of view, this second phase is marked by a decline of $cvpO_2$ whose values reache the threshold of begining hypoxia as defined by j. Grote et al (1971). The significant reduction in CBF tends to be compensated by an increase in both OER and GlucER as evidenced by increased (A-V) glucose and (A-V) oxygen. Nevertheless, $CMRO_2$ declines revealing the occurrence of a cerebral metabolic failure. The cerebral production of lactate does not increase, a fact which expresses that the glucose taken by the brain is not metabolized but is rather hydrated in astrocytes to favor edema (table II).

Table II : Cerebral hemodynamics and metabolism over the second phase of evolution

Parameters	Time (min) post-injection			
	30 (N=14)	60 (N=11)	90 (N=8)	120 (N=14)
CBF ($ml.min^{-1}.100\ g^{-1}$)	161 ± 23.7	113 ± 19.5 *	92 ± 18.9 *	45 ± 5.5 *
Perfusion pressure (kPa)	19.7 ± 0.49	18.5 ± 0.58 *	17.9 ± 0.91 *	16.0 ± 0.77 *
CVR	157 ± 23.3	227 ± 47.4 *	245 ± 34.0 *	490 ± 97.2 *
$cvpO_2$ (kPa)	5.29 ± 0.250	5.08 ± 0.320 *	4.45 ± 0.330 *	3.54 ± 0.050 *
(A-V) QO_2 ($ml.l^{-1}$)	44 ± 5.3	48 ± 5.4 *	79 ± 13.3 *	91 ± 7.5 *
OER (%)	26 ± 2.5	29 ± 3.3 *	38 ± 4.4 *	54 ± 1.7 *
$CMRO_2$ ($ml.min^{-1}$)	6.1 ± 0.61	4.7 ± 0.48 *	5.7 ± 0.62	4.4 ± 0.54 *
(A-V) gluc. ($mmol.l^{-1}$)	0.36 ± 0.072	0.43 ± 0.110	0.76 ± 0.153 *	1.00 ± 0.157 *
GlucER (%)	7.7 ± 1.3	10 ± 2.6	17 ± 3.6 *	21 ± 2.7 *
(A-V) Lact. ($mmol.l^{-1}$)	- 0.02 ± 0.071	- 0.25 ± 0.099 *	- 0.30 ± 0.126	- 0.412 ± 0.144 *

* $P \leq 0.05$ according to Student's t test or to Cochran's t test with regards to time 30 min.
N = number of data

The third period corresponds to an hemodynamic and metabolic failure ; it starts about 2 hours post-injection of the clot and has been followed up for 2 hours.

CBF drops dramatically so as to reach less than 10 percent of the initial value at the end of the experiment (fig 1), this drop not being originated by the reduction in MABP since its values do not bring down below 9 kPa ; consequently, the perfusion pressure is significantly reduced (fig 2).

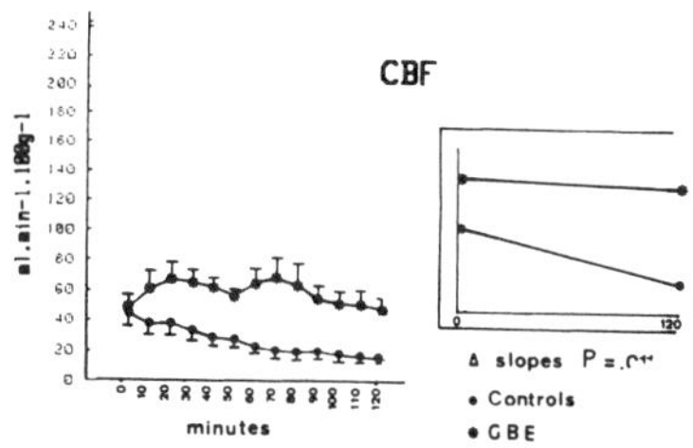

Fig. 1. Comparative evolution of cerebral blood flow

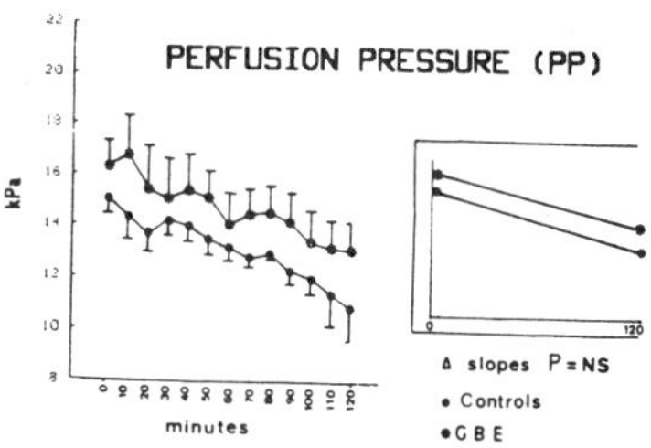

Fig. 2. Comparative evolution of perfusion pressure

So it is not surprising that CVR increases by more than 100 percent (fig 3)

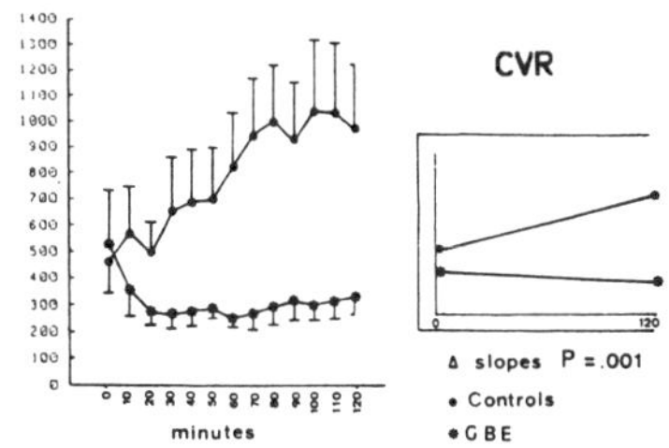

Fig. 3. Comparative evolution of cerebrovascular resistance

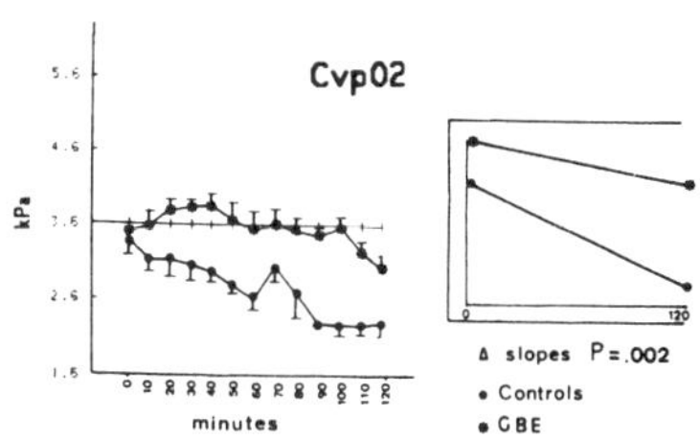

Fig. 4. Comparative evolution of cerebral venous pO_2

The oxygen delivery to the brain is impaired ; $CMRO_2$ is significantly lowered (fig 6) ; the capillary oxygen which illustrates the lenght of the arteriolar capillary along which the exchanges in O_2 occur is enlarged showing the difficulty for O_2 to reach the brain tissue (fig 5) ; all these events end in a reduction of $cvpO_2$ to the lethal zone as defined by J. Grote (fig 4) ; in conclusion, the oxygen availability is markedly impaired.

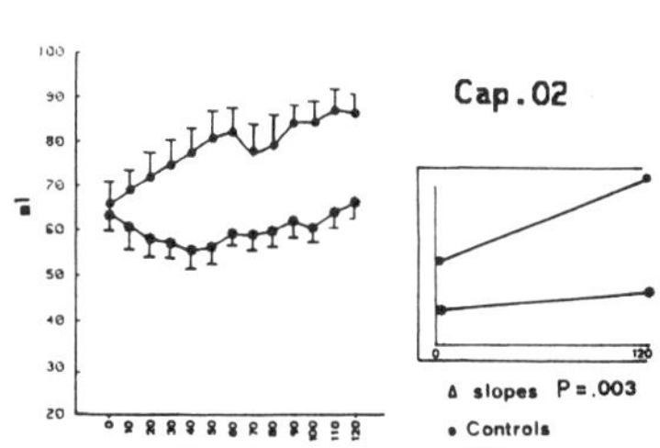

Fig. 5. Comparative evolution of capillary oxygen

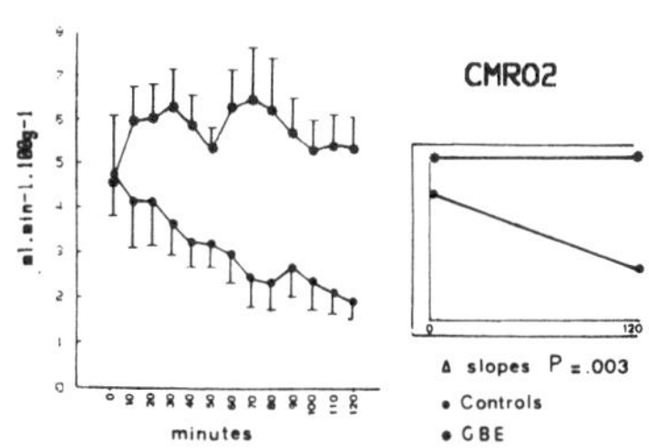

Fig. 6. Comparative evolution of cerebral metabolic rate in O_2

In parallel, CMR glucose increases (fig 7), the final result being a poor oxidation of this metabolite as evidenced by the significant increase of G/O ratio (fig 8). This is confirmed by an enhancement of the lactate production by the brain (fig 9).

Finally, it must be mentionned that 40 percent of dogs die prior to the end of the second hour of evolution.

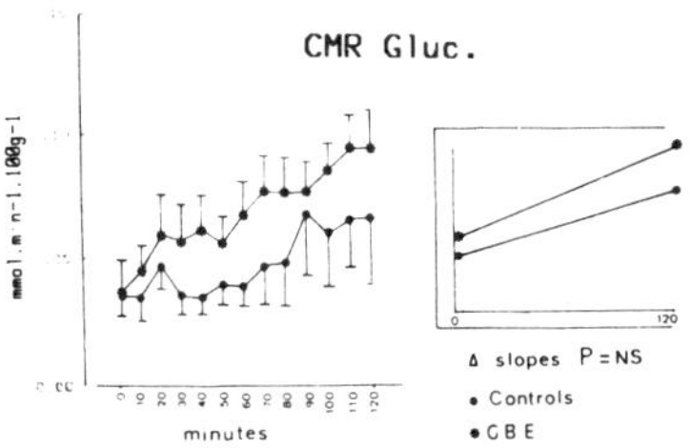

Fig. 7. Comparative evolution of cerebral metabolic rate in glucose

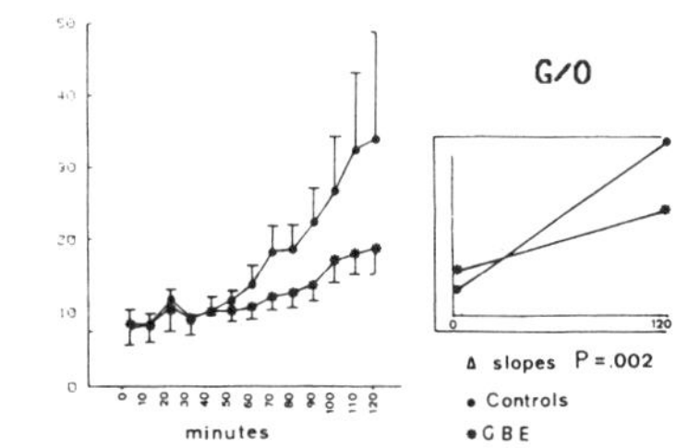

Fig. 8. Comparative evolution of G/O ratio

The activity of GBE has been investigated over this last period. CBF increases from the begining of GBE perfusion and remain significantly higher than in the controls (fig 1) ; nevertheless, a slight drop in perfusion pressure is similar in both controls and GBE series (fig 2) although CVR not only does not increase but moreover declines in comparison to what was observed 2 hours after the injection of the clot (fig 3).

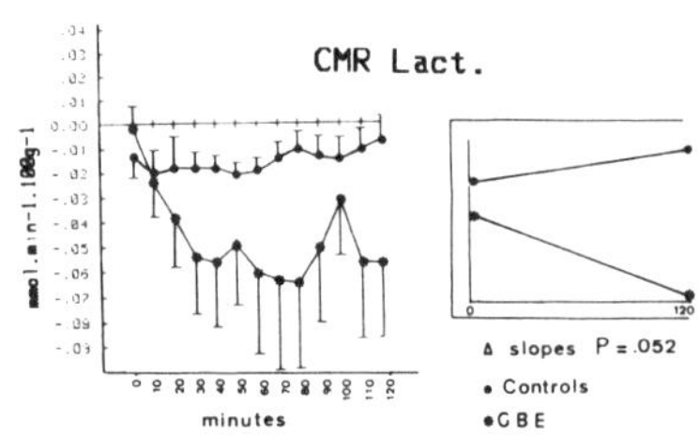

Fig. 9. Comparative evolution of cerebral metabolic rate in lactate

The oxygen availability is improved since $CMRO_2$ is preserved and even slightly increased (fig 6) ; the capillary O_2 does not increase significantly (fig 5) and the $cvpO_2$ remains, for at least 100 min, above the threshold of begining hypoxia (3.5 kPa) (fig 4). Although CMR glucose raises (fig 7), $CMRO_2$ is high enough as to allow a correct oxidation of the glucose taken by the brain as illustrated by the evolution of the G/O ratio which does not increase in the same proportion as in the controls (fig 8) ; nevertheless the cerebral production in lactate is increased (fig 9). No exitus has been observed in this treated group.

CONCLUSION

Cerebral infarction is one of the major brain injury. Ischemia resulting from embolisms is concomitant with reactive hyperhemia and finally ends in hemodynamic and metabolic failures.

GBE perfused from the begining of the latter can protect the brain in that CBF stops its decline and even slightly increases, $cvpO_2$ remained for a long time above the hypoxic threshold, $CMRO_2$ is maintained allowing the oxidation of the glucose taken by the brain.

All these facts lead to the conclusion that GBE may be helpful in the reanimation of the brain in case of acute ischemic injury.

REFERENCES

Grote, J. ; Kreuscher, H. ; Schubert, R., Ross, H.J. (1971) : Investigations on the influence of paO_2 and $paCO_2$ on the regulation of cerebral blood flow in dogs. In : Brain and Blood Flow. R.W. Ross Russel ed., Pitman Medical and Scientific Publishing Co. Ltd. pp. 200-204.

Lassen, N.A. (1966) : The luxury perfusion syndrome and its possible relation to acute metabolic acidosis localized within the brain. Lancet 2 : 1113-1115.

Olsen, T.S. ; Larsen, B. ; Skiver, E.B. ; Herning, M. , Enevoldsen, E and Lassen, N.A. (1981) : Focal cerebral hyperhemia in acute stroke. Incidence, Pathophysiology and clinical significance. Stroke 12, 5 : 598-607.

Trinder, P. (1966). Ann. Clin. Biochem. 6 : 24.

AKNOWLEDGEMENTS

The author thanks IPSEN Laboratories for having provided Ginkgo biloba extract. He also expresses his thanks to Mr Jean Angignard, Mrs Dominique Angignard and Suzanne Weber for their technical assistance.

Effects of Ginkgo Biloba Extract on Organic Cerebral Impairment. A. Agnoli, J. R. Rapin, V. Scapagnini, W. V. Weitbrecht. John Libbey Eurotext Ltd © 1985

Effects of Ginkgo biloba extract on two types of cerebral edema

M. G. Borzeix

Département de Thérapeutique Expérimentale, SIR International, 6 rue Blanche, 92120 Montrouge, France

SUMMARY

Whatever its origin is toxic, traumatic, ischemic, etc, edema is the common reaction of the brain tissue to injury. Triethyltin (TET) poisoning in the Rat results in a marked neurologic deficit which derived from a splitting of myelin bleds due to fluid accumulation. On another hand, a rough disruption of the dura-mater induces in the Rabbit, deep disturbances of the electrical activity of the brain recorded in the surroundings of the injury. The administration of Ginkgo biloba (GBE) together with TET, completely prevents in the Rat the occurrence of edema and consequently of the neurologic deficit. Similarly, the ingestion of GBE to rabbits 18 hours after the trauma-induced cerebral injury produces a marked improvement of the electrical activity of the brain mainly due to a decline in the relative proportion of slow waves.

KEY WORDS

Triethyltin edema, traumatic edema, rats, neurologic deficit, rabbits, Na+, K+, water, EEG, Ginkgo biloba extract.

INTRODUCTION

Various alkyl tin and lead compounds are known to produce selective edema of the Central Nervous System among which triethyltin (TET) is the more potent (P.N. Magee et al, 1957). Neurologic signs develop within one week, starting with paralysis of the hind legs which is followed by a marked loss of weight, diarrhea, apathy, protrusion of the eyeballs, etc ... In white matter, an accumulation of fluid within myelin blebs appears as reported by F.P. Aleu et al (1963) ; the water content of tissue, referred to dry weight, markedly increases as does the sodium content whereas the potassium amount declines. In grey matter, few changes occur : swelling of astrocytic processes can be observed on electron micrographs along with a slight increase in water content (L. Bakay, 1965 ; R. Katzman, 1967) ; but there is no breakdown of the blood-brain-barrier. Although the mechanism of TET still remains unknown, the development of TET-induced edema can be prevented by drugs acting on electrolyte movements in the CNS, certainly through an effect on ATPase activity (R.M. Torack et al 1960).

Another type of cerebral edema which mainly involves the gray matter is a trauma-induced cerebral edema which results from a rought disruption of the dura-mater (M.G. Borzeix et al 1972). This produces deep disturbances of the electrical activity in the surrounding areas, changes which can be computerized in terms of energy power for the different bands of frequency.

The goal of this study was to investigate the anti-edematous activity of Ginkgo biloba extract (GBE) through its effect upon the two above-mentioned experimental models of cerebral edema.

MATERIAL AND METHODS

Triethyltin-induced edema in the Rat.

This study has been undertaken on 75 male SPF Sprague-Dawley rats, weighing about 220 g at the begining of the experiment ; they were given an aqueous solution of TET chloride (0.002 percent) as sole drinking water, for 2 weeks. The neurologic deficit was estimated first, by the mean angle of an inclined plane on which the rat can stay on, and second, by the gait, whatever it was normal (index 10), on point of paws (index 8), hypertonic with enlargement of the polygone of support (index 2) or finally the rat exhibited a lateral decubitus (index 0) ; this estimation has been performed at day 10 and day 15 of the TET intoxication. Then the rats were killed by decapitation ; the brain rapidly removed was placed for 72 hours in a dry hot-air bath in order to calculate the water content (%) and then to allow the determination of Na^+ and K^+ amounts on the dry residue ($mmol.kg^{-1}$ calculated to dry weight).

Rat were given the Ginkgo biloba extract by oral route, all along the experiment, at 3 daily doses, respectively 10 - 30 and 90 $mg.kg^{-1}$; the effect of GBE was compared to the values obtained in a group of TET-poisonned controls and on a series of healthy rats, each group involving 15 animals.

Trauma-induced cerebral edema in the Rabbit.

In 40 male rabbits, weighing about 2.5 kg, a hole (10 mm diameter) was performed into the skull, over the parieto-occipital area. Twenty four hours later, the dura-mater was roughly disrupted inducing thus an immediate edematous reaction of the nervous tissue. The electrocorticogram (ECoG) was continuously recorded from 4 steel electrodes screwed into the skull in the surroundings of the lesioned zone as well as on the controlateral hemisphere (M.G. Borzeix et al, 1972). The ECoG was then split in four bands of frequency by means of a filter system (Spectrovar, Alvar Electronic) and digitalized (Integrator, Alvar Electronic) in order to evaluate the energy power of each band expressed in percentage of the total energy power ; the 4 bands of frequency were respectively : delta : 0.5-3 Hz ; theta : 4-9 Hz ; alpha : 10-17 Hz and beta : 18-30 Hz.

Three series of ten rabbits each, were given GBE 18 hours after the disruption of the dura-mater by oral route, at the respective doses of 12.5, 25 and 50 mg.kg-1. The results obtained were compared to the reference values (before treatment).

COMMENTS

TET-induced cerebral edema in the Rat.

The first sign to be noted in the controls is a marked loss of bodyweight around the 10th day of TET poisoning ; at the same time, a paralysis of the hind legs starts which is responsible for the reduction of the neurologic indices, that being reinforced at the end of the second week of TET diet (table I).

The electrolyte content calculated to the dry weight of tissue shows a significant increase of sodium in edematous tissue whereas potassium does not change significantly (table I).

According to those facts, it appears clearly that the oral administration of GBE significantly prevents the occurrence of the cerebral edema and of its related neurologic signs (table I).

Table I : Electrolyte content of rat brain and neurologic scores

Series	Daily Dose mg.kg^{-1} P.O	N	Cerebral content %	mmol.kg^{-1} dw		Neurologic index		Inclined plane (degree)	
			H_2O	Na^+	K^+	D 10	D 15	D 10	D 15
Controls	-	15	78.3 ± 0.10	201 ± 1.8	382 ± 7.7	10	10	90	90
TET-poisoned rats	-	15	80.5 ± 0.17 (°)	290 ± 5.3 (°)	373 ± 8.2	4.1 (°)	0 (°)	30 (°)	0 (°)
GBE-treated rats	10	15	79.7 ± 0.18 *	249 ± 6.7 *	372 ± 4.9	9.7 *	9.7 *	90 *	90 *
	30	15	78.6 ± 0.14 *	209 ± 3.5 *	377 ± 6.7	10 *	10 *	90 *	90 *
	90	15	78.4 ± 0.07 *	202 ± 2.4 *	390 ± 4.3	10 *	10 *	90 *	90 *

dw = dry weight
D 10 - D 15 = day 10, day 15
$P \leqslant 0.001$
(°) versus controls * versus TET-poisoned rats

The effect of GBE is highly significant, whatever the dose used, from 10 to 90 mg.kg-1 in that no edema is observed and consequently no neurologic deficit develops (table I).

Trauma-induced cerebral edema in the Rabbit.

EcoG disturbances deriving from edema are characterized on fig 1.

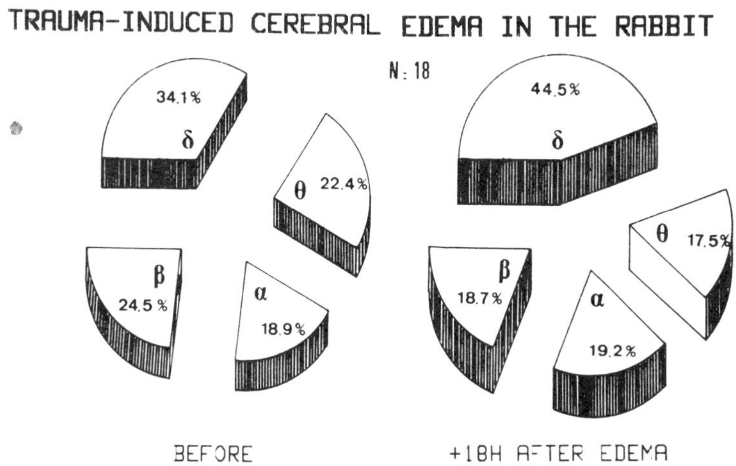

Fig. 1. : Electrocorticographic disturbances related to trauma-induced cerebral edema in the Rabbit.

The main fact is a tremendous increment of the proportion of slow waves (delta rhythm) whereas in the mean time, the relation percentage of θ and β frequencies declines ; these changes are responsible for a drop in the values of $\frac{\theta + \beta}{\delta + \alpha}$ ratio which illustrate the proportion of physiological rhythms (θ and β) recorded in posterior areas, in the conscious rabbit and pathologic ones (δ and α) ; after trauma-induced injury, the former decline whereas the latter predominate. For instance in 18 healthy rabbits the average value $\frac{\theta + \beta}{\delta + \alpha}$ ratio (x 100) is 94 ± 8.0. The evolution of this ratio of accounts for an impairment (increase) or an improvement (decrease) of the electrogenesis. Particularly, it shows that GBE, at 50 $mg.kg^{-1}$ significantly improves the aspect of ECoG that being mainly related to a reduction of slow waves (table II).

Table II : Evolution of the $\frac{\theta + \beta}{\delta + \alpha}$ ratio (x 100) over the 6 hours following GBE treatment

Time post-treatment (hour)		Controls N=10	GBE (mg.kg^{-1} P.O.) 12.5 N=10	25 N=10	50 N=10
Reference value		51 $\pm$ 4.6	43 $\pm$ 5.2	44 $\pm$ 3.9	38 $\pm$ 2.8
Δ versus reference value	1	- 3 $\pm$ 3.0	- 1 $\pm$ 3.8	+ 3 $\pm$ 3.3	+ 7 $\pm$ 3.8 (*)
	2	- 4 $\pm$ 3.1	- 1 $\pm$ 3.1	+ 2 $\pm$ 4.4	+ 5 $\pm$ 3.9
	3	- 1 $\pm$ 2.8	+ 3 $\pm$ 4.3	- 1 $\pm$ 3.8	+ 8 $\pm$ 3.5 (*)
	4	- 6 $\pm$ 3.1	+ 2 $\pm$ 4.9	0 $\pm$ 3.3	+ 12 $\pm$ 5.6 (*)
	5	- 2 $\pm$ 3.1	+ 3 $\pm$ 4.4	+ 1 $\pm$ 3.7	+ 10 $\pm$ 3.4 (*)
	6	- 4 $\pm$ 3.6	+ 4 $\pm$ 3.8	- 1 $\pm$ 3.3	+ 12 $\pm$ 4.9 (*)

(*) $P \leqslant 0.05$ according to paired t test

CONCLUSION

Two experimental models of cerebral edema have been considered.

Triethyltin poisoning in the Rat provokes the occurrence of an edema of the brain tissue mainly located to the white matter, the related neurologic deficit deriving from the accumulation of fluid within the myelyn blebs. Ginkgo biloba extract given to rats together with TET completely impedes the occurrence of any edematous reaction and consequently of any neurologic deficit ; this therapeutic effect is observed similarly at 10, 30 and 90 mg.kg^{-1} P.O.

The rough disruption of the dura-mater induces in the Rabbit a reactive edematous reaction of the cerebral tissue which is characterized by deep disturbances of the EEG in the surrounding areas. The oral administration of GBE (50 mg.kg^{-1} P.O.) results in an improvement of the EEG particularly through a reduction of the proportion of slow waves.

Those results could indicate that one of the main therapeutic properties of GBE is the potent anti-edematous effect that is observed in 2 different experimental conditions, one involving a cytotoxic effect and the other being originated by a vasogenic reaction which is very often observed after traumatic insult or during neurosurgery.

REFERENCES

Aleu, F.P. ; Katzman, R. and Terry, R.D. (1963) : Fine structure and electrolyte analysis of cerebral edema induced by alkyl tin intoxication. J. Neuropath. Exp. Neurol. 22 : 403-413.

Bakay, L. (1965) : Morphological and chemical studies in cerebral edema : triethyltin induced edema. J. of Neurological Sciences 2 : 52-67.

Bakay, L. and Lee, J.C. (1965). In : Cerebral Edema, Ed. L. Bakay and J.C. Lee, C. Thomas Publish.

Borzeix, M.G. ; Labos, M. and Cahn, J. (1972). A propos d'un modèle expérimental d'oedème cérébral pour l'étude des substances agissant sur le débit sanguin et/ou le métabolisme cérébral. Agressologie 13, 4 : 257-260.

Borzeix, M.G. ; Weber, S. and Akimjak, J.P. (1978) : Cerebral edema induced by triethyltin (TET) in the rabbit and in the rat. In : Proc. of the 7th international Congress of Pharmacology - Satellite Symposium of Reims.

Katzman, R. ; Aleu, F.P. and Wilson, C. (1963) : Further observations on triethyltin edema. Arch. Neurol 9 : 178-187.

Katzman, R. (1967) : Biochemical correlates of cerebral edema - In Brain edema : pp. 461-467, Ed. I. Klatzo and F. Seitelberger Springer Verlag Publish.

Magee, P.N. ; Stoner, H.B. and Barnes, J.M. (1957) : The experimental production of edema in the central nervous system of the rat by triethyltin compounds. J. Path. Bact. 73 : 107-124.

Torack, R.M. ; Terry, R.D. ; Zimmerman, H.M. (1960) : The fine structure of cerebral fluid accumulation - II. Swelling produced by triethyltin compounds and its comparison with that in the human brain. Am. J. Path. 36 : 273-287.

AKNOWLEDGEMENTS

The author thanks IPSEN Laboratories for having provide Ginkgo biloba extract. She also thanks particularly Messrs. J.P. Akimjak, C. Hartl and Mrs. D. Angignard for their technical assistance.

Effects of Ginkgo Biloba Extract on Organic Cerebral Impairment. A. Agnoli, J. R. Rapin, V. Scapagnini, W. V. Weitbrecht. John Libbey

Hemorheological effects of standardized Ginkgo extract in vitro

E. Ernst

Hemorheology Research Unit, Clinic of Physical Medicine, University of Munich, Germany

ABSTRACT

Heparinized venous blood from 20 patients suffering from circulary disorders associated with abnormal blood rheology was incubated with standardized Ginkgo extract at various concentrations. Red cell filterability and red cell aggregation were determined. Results show a marked reduction in aggregation while filterability remained constant. The potential importance of the findings in the treatment of cerebro-vascular disease are discussed.

KEYWORDS

Ginkgo Biloba extract - blood rheology - in vitro study - red cell flexibility - red cell aggregation.

INTRODUCTION

Blood fluidity is determined mainly by hematocrit, plasma viscosity and red cell microrheology (i.e. aggregation and flexibility)(1). There is abundant evidence for blood fluidity being one of several determinants of cerebral blood flow (CBF). In polycytemia of various etiologies (=high blood viscosity) CBF is low (2 - 5) while in anemia (=low blood viscosity) it is high (2,6). Experimental hematocrit modification results in an inverse alteration of CBF (5, 6, 7, 8). Hemodilution in patients suffering from cerebrovascular disease (CVD) yields symptomatic improvement (9) and increases cerebral oxygen flow (10). Furthermore high plasma viscosity has been shown to be associated with low CBF(11-13) and its normalisation leads to increased flow (9).

These data suggest that normalisation of pathological blood rheology like it exists in CVD (2) can be used in therapeuthic. The aim of the present communication was therefore to investigate whether a standardized Ginkgo extract as being used with symptomatic success in CVD (14) is hemorheologically active in vitro.

METHODS

Venous blood was drawn from 20 patients suffering from circulatory disorders known to be associated with abnormal red cell rheology (cerebral and peripheral vascular disease) and anticoagulated with 12,5 U/ml Lithium Heparin. All blood samples included in this study showed abnormal red cell filterability and aggregation . Blood samples were incubated with 0, 10, 20, 50 µl/ml standardized Ginkgo extract at 37°C for 1 hour. Measurements were executed in triplicate in a single blind fashion. Red cell filterability was performed by resuspending erythrocytes in autologous plasma at a hematocrit of 5% and filtering this suspension through 5 µm pores (Nucleopore) with a pressure head starting at 1 cm H_2O decreasing to zero. The number of cells having passed the filter within 1 minute in relation to the ones that could have passed, gives the filterability index (F.I.) which is a measure for red cell flexibility. (15).

Red cell aggregation was assessed in an automated cone and plate rotational viscometer, where the sample is sheared to disperse red cells. Thereafter the sample is at rest, hence red cells can aggregate.This process alters the light transmission through the sample. Light transmission is integrated over time (60 s), giving a measure for the speed and extend of red cell aggregation (16).

Results were analyzed using the two sided t-test (student). The null hypothesis was rejected when $p < 0,05$.

RESULTS

Red cell filterability was not altered by Ginkgo at any of the concentrations employed (Table I., Fig. 1).

		Comparison to baseline value	
Concentration	$\bar{x}$ (Fig.I)	$\bar{x}$	p
0	0.51	+ 0.01	
10 µl/ml	0.52	+ 0.01	0.2787
20 µl/ml	0.52	+ 0.01	0.4040
50 µl/ml	0.49	- 0.02	0.2065

Table I: Red cell filterability results with various concentration of standardized Ginkgo extract (high FI means good flexibility ; normal value for FI = 0,68 ± 0,11).

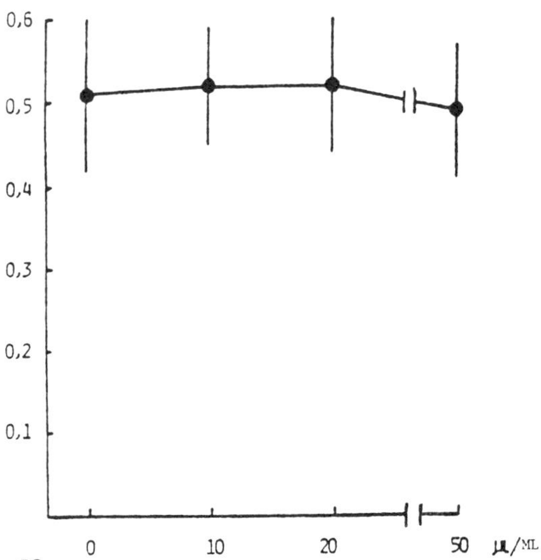

Fig. 1 : Graphic outline of Ginkgo-induced changes on red cell filterability.
FI vertical axis, Ginkgo concentration horizontal axis.

Red cell aggregation was reduced significantly at concentrations 10 and 20 µl/ml. This effect was abolished at the highest concentration of 50 µl/ml (Table II, Fig. 2). Within the therapeutical range (10 - 20 µl/ml) there seems to be a dose dependent reduction of red cell aggregation by Ginkgo.

Concentration	$\bar{x}$	SD	Comparison to baseline value $\bar{x}$	SD	p
0	17.7	2.6			
10 µl/ml	15.0	2.7	- 2.7	3.2	0.0012
20 µl/ml	13.6	2.5	- 4.1	2.3	<0.000001
50 µl/ml	16.5	4.7	- 1.2	3.5	0.15

Table II: Red cell aggregation results with various concentration of standardized Ginkgo extract (high values for aggregation mean high aggregation tendency; normal value = 10,3 ± 2,4).

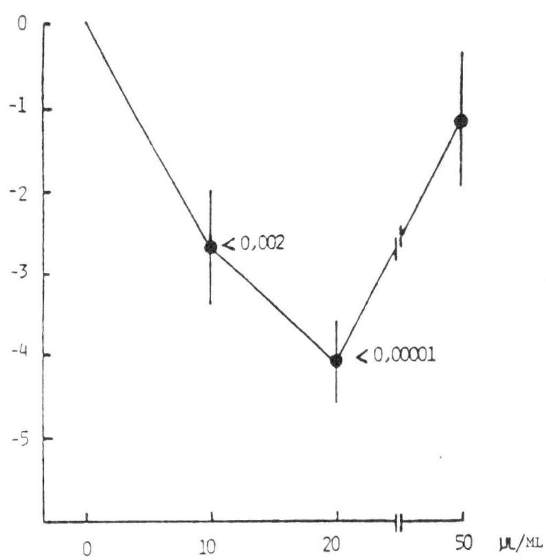

Fig. 2 : Graphic outline of Ginkgo-induced changes on red cell aggregation. Vertical axis change in absolute units, horizontal axis Ginkgo concentration.

DISCUSSION

Hemorheological effects of Ginkgo have been reported in experimental animals (17), in volunteers as well as in patients with circulatory disorders (18) and in those with complete stroke (19). In vitro experiments in this yield field can give valuable information about the specific mode of rheological action. The present findings suggest a beneficial effect upon blood rheology through decreasing red cell aggregation, which is known to be one cause for blood's fluidity at low shear stresses (1).

To express complex phenomena briefly, red cells can form two - and three - dimensional networks in vivo and in vitro, a process which is fully reversible. Aggregation is induced by plasmatic macroproteins (i.e. fibrinogen, IgM, (x_2 macroglobulin), which bind to the membranes of two cells.
In this situation the binding forces overcome the desaggregating forces (electrostatic repulsion, membrane deformation and shear stress) (Fig.3). Hence red cell aggregation takes place only when shear stress is low, or, in circulatory terms, when stasis or prestasis occurs (20).

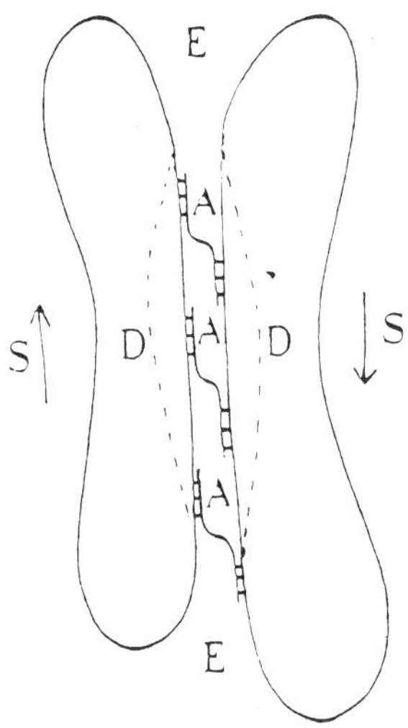

Fig. 3 : Schematic drawing of the red cell aggregation phenomenon and the forces involved. Aggregation (A) are in balance with desaggregating forces (E = electrostatic repulsion, D = membrane deformation, S = shear stress).

It has been observed that under certain circumstances red cell aggregation can be a major hindrance for the circulation (21).
Thus enhancing the fluidity of blood by reducing red cell aggregation seems a plausible principle for increasing blood flow, especially in low flow areas. The therapeutic use of Ginkgo under this aspect might be promissing in a number of low flow states, one of which is CVD.

CONCLUSION

It is always problematic to extrapolate conclusions from in vitro experiments (which have a value in their own rights) to any in vivo situation. Hence the present findings need to be tested in clinical trials. If results would be similary promissing, one could postulate that Ginkgo's clinical efficacy in CVD (22, 23) is partly brought about by rheological mechanism (s). This would be yet another confirmation of blood rheology being of crucial importance in CVD, a thesis which finds a strongest support in a number of epidemiological, prospective and retrospective clinical studies (24 - 27).

REFERENCES

1. Chien S.(1975) : Biophysical behavior of red cells in suspension. In : The Red Blood Cell. Vol. II, Academic Press, San Francisco, pp. 1031 - 1131.
2. Gottstein U.(1965) : Physiologie und Pathophysiologie des Hirnkreislaufs. Med. Welt 15, 715.
3. Nelson D., Fazehas J.F.(1956) : Cerebral blood flow in polycythemia vera. Arch. Int. Med. 98, 328.
4. O'Brien M.D.(1971) : Some neurological aspects of dementia. Gerontologia Clinica 13, 339.
5. Thomas D.J. et al (1977) : The influence of hematocrit on cerebral blood flow in man. Lancet, II, 941.
6. Scheinberg P.(1951) : Cerebral blood flow and metabolism in pernicious anemia. Blood 6, 213.
7. Thomas D.J. et al (1979) : Prevention of stroke - the viscosity factor. Cerebral Vascular disease ; proc. 9th int. Salzburg Conf. Meyer J.S.,Lechner H., Reivich M. (Eds). Excerpta Medica pp. 211 - 215.
8. Paulson O.B., Henriksen L;, Smith R.J.(1979) : The effect of hemodilution on cerebral blood flow and blood gases in patients with polycythemia. Acta neurol. Scanda. 60, suppl. 72, 588.
9. Thomas D.J.(1979) : The influence of blood viscosity on cerebral blood flow and symptoms. in : Progress in Stroke Research. Greenhalgh R.M., Rose F.C. (eds). Tunbridge Wells, Pitman Medical pp. 47 - 55.
10. Wade J.P.(1981) : Studies on the relative importance of blood viscosity and oxygen carriage in determining cerebral blood flow in primary and secondary polycythemia. MD Thesis, University of Cambridge.
11. Humphrey P.R.D. et al (1980) : Viscosity, cerebral blood flow and hematocrit in patients with paraproteinemia. Acta neurol. Scand. 61, 201.
12. Bown M.M., Pearson T.C., Marshall M.(1982) : Effects on changes in plasma viscosity on cerebral blood flow in man. Clin. Hemorheol. 2, 5/6.
13. Grotta J. et al (1982) : Whole blood viscosity parameters and cerebral blood flow. Stroke 13.
14. Heiss W.D., Zeiler K.(1978) : Medikamentöse Beeinflussung der Hirndurchblutung. Pharmakotherapie 1, 3, 137.
15. Dodds A.J., Dormandy J. et al (1976) : Haemorheological response to plasma exchange in Raynaud's Syndrome. Brit. Med. J.2, 1186.
16. Schmid-Schönbein H. et al (1982) : New hemorheological techniques for the routine laboratory. Clinic. Hemorheol. 2, 1/2, 93.
17. Gabard B., Chatterjee S.S. : Personal communication.
18. Hossman W., Heiss W.D., Bewermeyer H.(1980) : Fortschritte der technischen medizin in der neurologischen Diagnose und Therapie. Wien pp. 445.
19. Witte S., Anadere I., Chmiel H.(1983) : The influence of a Ginkgo biloba extract on the increased viscoelasticity of blood by complete stroke. Clin. Hemorheol.3,3, 291.
20. Chien S. et al (1970) : Shear dependent interaction of plasma proteins with erythrocytes in blood rheology. Am. J. Physiol. 219, 143.
21. Appelgreen L.(1982) : Perfusion and diffusion in shock. Acta Physiol. Scand. Suppl.378.
22. Haan J. et al (1982) : Ginkgobiloba Flavonglycoside - Therapiemöglichkeit der zerebralen Insuffizienz. Med. Welt 27, 2.
23. Eckmann F., Schlag H.(1982) : Kontrollierte Doppelbliendstudie zum Wirkungsnachweis von Tebonin forte bei Patienten mit zerebrovaskulärer Insuffizienz. Fortschr.Med. 31, 3.
24. Kannel W.B. et al (1972) : Hemoglobin and the risk of cerebral infarction. The

Framingham Study. Stroke 3, 4, 409.
25. Meade J.W. et al (1980) : Hemostatic function and cardiovascular death: Early results of a prospective study. Lancet 1, 2, 1050.
26. Harrison M.J.G. et al (1981) : Effect of hematocrit on carotid stenosis and cerebral infarction. Lancet. 2, 114.
27. Lowe G.D.O. et al (1983) : Relation of artrial fibrillation and high hematocrit in acute stroke. Lancet 1, 784.

Effects of Ginkgo Biloba Extract on Organic Cerebral Impairment. A. Agnoli, J. R. Rapin, V. Scapagnini, W. V. Weitbrecht. John Libbey

Multiplicity of effects of Ginkgo biloba extract: currents status and new trends

K. Drieu

I.H.B.-IPSEN, 30 rue Cambronne, 75015 Paris, France

ABSTRACT

The chemical complexity of Ginkgo biloba extract prepared from dry leaves of the Ginkgo tree, leads to a multiplicity of effects. Pharmacological studies in which GBE was effective are reviewed. The therapeutic benefit of GBE is supported by - hemodynamic properties leading to the equilibrium of the blood flow in small vessels - normalization of metabolic parameters in ischemic and hypoxic models - and restoration of muscarinic receptors in aged animals.
GBE has been shown in vitro and in vivo to be a free-radical scavenger; thus it is postulated that at least part of its peculiar qualities are due to a "membrane effect", some others being related to enzymes and neuromediators regulation.

KEYWORDS

Ginkgo biloba extract - pharmacological studies review - brain metabolism - membrane - free radicals - ischemia - hypoxia - neuromediators -

INTRODUCTION

The Ginkgo biloba extract is a defined and complex product prepared from the dry green leaves of Ginkgo biloba. The leaves are harvested from trees growing in plantations in South Corea and Japan.
The crude extract is purified and enriched during an extraction process consisting in fifteen different steps which leads to a precisely standardized extract.
It consists of about 50 % flavonoïdic substances: kaempferol, quercetine and isorhamnetine glucosides, some of them specific to Ginkgo, and proanthocyanidines.

Fig. 1 :

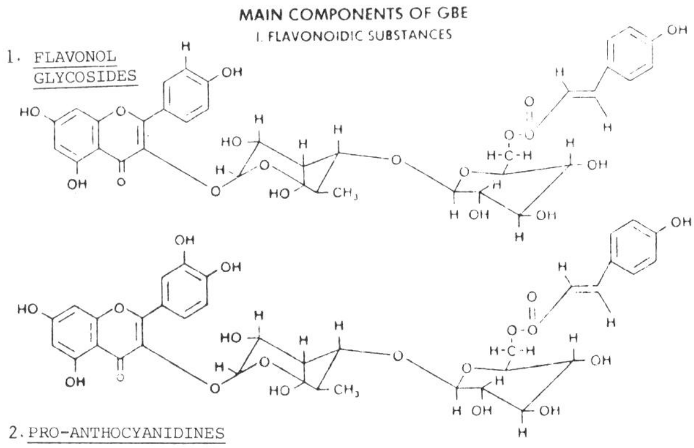

The non-flavonoïdic part of the extract consists of organic acids and terpenoïds, which are characteristic of Ginkgo biloba: they are the ginkgolides and the bilobalide.

Fig. 2 :

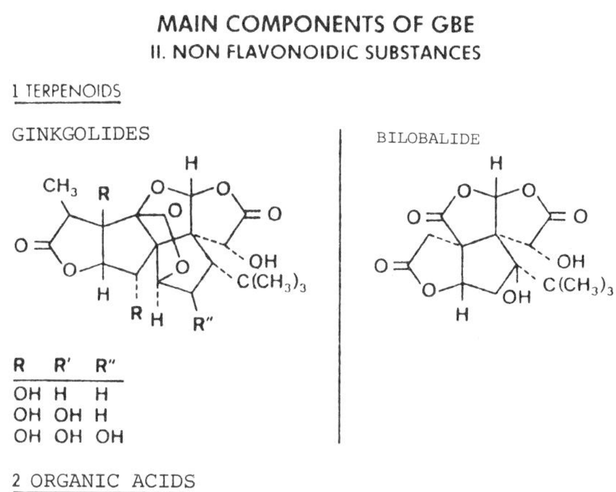

The chemical complexity of the extract brings about a multiplicity of effects which are interdependant on each others. In fact the activity and the mechanisms of action of GBE form a kind of puzzle: piece by piece the image of Ginkgo is being built. A fractionation of the total extract leads to the separation of the different groups of substances and also pure substances: on most experiments in which the total extract is active, the fractions or single compounds did not give the whole activity of the total extract.

Advances in the knowledge of the brain and of its pathology, the development of pathological models, the progress of technology helped to show the activity of GBE in experimental studies and from the beneficial effects observed in animals and man we went on to examine the mechanisms of action.

At the begining we observed the protective effect of GBE under hypoxia. It increases dose-dependently the survival time of mice under hypobaric hypoxia (1).

It brings back to normal the deoxyglucose uptake in mouse brain under normobaric hypoxia without markedly affecting the cerebral blood flow (2).

In different models of brain edema, GBE reduces the formation of the edema and its consequences. In a model of quantified embolization by labelled microspheres, GBE gave an inhibition of the edema (3) accompanied by restored ATP and glucose levels in the ischemic hemisphere, together with an equilibration of the cerebral blood flow.

Removing of the dura-mater in rabbits induces a traumatic edema and thus, disturbances of the ECoG: GBE restores dose-dependantly the ECoG (4).

GBE was also studied on cerebral infarction induced by arachidonic acid in rats (5) and dogs (6) or by intracarotid injection of an autologous clot (7). In all these models GBE was effective.

Triethyltin edema is a model of choice of some scientific teams (8): the understanding of the fundamental mechanisms of TET-toxicity may help to explain why and how some active substances in cerebral insufficiencies are acting. GBE is certainly one of the most active in this model, as it not only prevents (9) but cures (10) the triethyltin edema.

J.R. RAPIN and co-workers (impublished data) showed also that GBE is able to decrease the permeability of the blood-brain-barrier to a small labelled polypeptide. This was observed after an hypertensive rise (Fig. 3) or after microspheres embolization before the formation of the edema.

Fig. 3 :

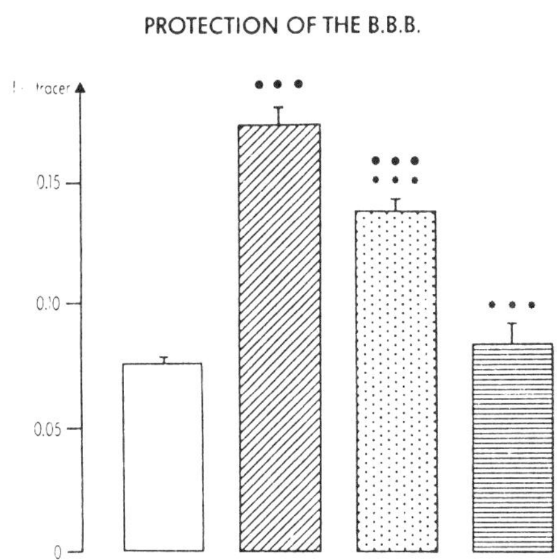

The improvement by GBE of metabolic parameters in ischemic models (3), of brain energy metabolism in hypoxia (11) and the protection of the blood-brain-barrier are consequences of membrane and/or enzymatic effects.

Hemodynamic effects are linked to receptors and neuromediator release in the vessel wall (12-13). Globaly, treatment with GBE leads to a restoration of the equilibrium of blood circulation in small blood vessels, whether this has to be reinforced or limited.

What was observed concerning cognitive, behavioral and learning function may be related to the effect of Ginkgo biloba on neuromediators and at the level of their receptors, but also may be linked to membrane effects. Preservation of the membrane integrity is certainly the key element for proper functioning of an organ. Without coming into the biochemistry of free radicals and of the lipoperoxidation of membranes, it is important to remind that lipid radicals which are generated by free radicals influence the fluidity and the permeability of cell membranes: these effects have repercussions on capillaries, cells and neurones . This happens (14) in case of ischemia, especially partial ischemia, in case of traumatic lesions, in case of edema, in atherosclerosis processes. And also it seems that the free radicals are widely involved in the disturbances due to aging.

Ginkgo biloba extract has been shown to be a free radical scavenger in vitro in several models.

TABLE I :

FREE-RADICAL SCAVENGING ACTIVITY
in vitro
GBE: ED50 ≐ 10^{-3} mg/ml

1. Inhibition of the FENTON REACTION (C. Deby et al.)
   - physical evaluation (ESR)
   - chemical evaluation
2. Inhibition of the adriamycin semiquinone formation (C. Deby et al.)
3. Inhibition of the γ-irradition induced depolymerisation of hyaluronie acid (C. Deby et al.)
4. Inhibition of the chemoluminescence of luminol (A. Etienne et al.)

Precise, well standardized and direct in vivo models to show free radicals are not yet really available. But some works, by showing the inhibition of malonaldehyde formation, the antagonism of adriamycin toxicity (15-16) give indirect evidence of free radical scavenging or antilipoperoxydant effects of GBE.

TABLE II :

FREE-RADICAL SCAVENGING
AND/OR ANTILIPOPEROXIDANT ACTIVITY
OF GBE *in vivo*

1. Antagonism of Adriamycin effects
   - protection against lethality (S. Chatterjee)
   - antagonism of local edema (A. Etienne et al.)
2. Protection against alloxan diabetitis (S. Chatterjee et al.)
3. Decrease of MDA formation in brain homogenales *ex vivo* (S. Chatterjee et al.)
4. Decrease of MDA formation in rats after Arachidonic acid injection (P. Braquet et al.)

Furthermore, the function of neuromediator's receptors depends upon the integrity of cell membrane. The normalization of the muscarinic receptors in aged rats showed by J. Taylor (this symposium) may be related to the improvement of the fluidity of the cell membrane as is the normalization of the metabolic exchanges resulting in a better functioning of enzymes in the cell respiratory process.

Work aimed at increasing our knowledge of GBE is always in progress.
As phosphodiesterase was shown a long time ago to be inhibited by flavonoïds and by GBE, new experiments have been performed; GBE was shown ex-vivo to inhibit cyclic AMP phosphodiesterase of rat brain (particulate enzyme) and studies are in progress to examine its relation to calcium and calmodulin which regulate phosphodiesterase activity. Investigations of mitochondrial function following carotid-ligation-induced ischemia in the gerbil are also in progress and preliminary results have shown that GBE prevents the uncoupling of oxidative phosphorylation which is the key process of cellular energy metabolism.
Very recently, we obtained very promising results with a model of vasospasm (in rabbits) induced by the exposure of the surface of the brain to autologous serum: GBE injected intravenously inhibited the vasospasm and one could see on a screen the re-enlargment of pial arteries and the reflow of blood into capillaries; the effect is dose-dependent and a single injection of GBE is still effective after several applications of serum over a period of one hour (S. Reusse: to be published).

CONCLUSION

Thus, in pharmacological studies, we have shown the beneficial effect of Ginkgo biloba extract. The mechanisms of action are being elucidated. A lot remains to be done, but it is clear that, even though a major part of the effect of GBE seems to be due to a restoration of membrane integrity, the multiplicity of its effects is due to its chemical complexity.
Even if we had contributed by making a purified leaf extract, Nature made an adequate natural blend for therapeutic benefit.

1. Chatterjee S.S., Gabard B. (1981) : Protective effect of an extract of Ginkgo biloba and other hydroxyl radical scavengers against hypoxia. Eithth International Congress of Pharmacology, Tokyo, Abstract 866.
2. Rapin J.R., Duterte D., Le Poncin-Lafitte M. (1980) : Hypoxie normobare et activité cérébrovasculaire chez la souris. J. Pharmacol, janvier-mars,11 p.101 (ed. Masson - Paris).
3. Le Poncin-Lafitte M.C., Rapin J. Rapin J.R. (1980) : Effects of Ginkgo biloba on changes induced by quantitative cerebral microembolization in rats. Arch Int Pharmacodyn 243: 236-244.
4. Borzeix M.G., Labos M. Hartl C. (1979) : Comparative spectral EEG analysis during cerebral injuries of traumatic ischemic or toxic origine in the unanesthetized rabbit.
Cerebral Blood Flow and Metabolism.Acta Neurologica Scandinavica supp. 72 volume 60 - Ed. Fumio Gotoh, Hajime Nagai Yoshiaki Tazaki (Japan).
5. Braquet P., Braquet M., Deby C. (1983) : Oxydative damages induced by cerebral ischemia: protective role of some radical scavengers and related drugs. Journal of cerebral blood flow and metabolism 1983 vol. 3, suppl. 1, 564-565.
6. Cahn J. Borzeix M.G., Angignard J. Cahn R. (1979) : Experimental multiple cerebral infarct produced in Mongrel dogs by perfusion of sodium lactate or of sodium arachidonate in the vertebral artery (51 Mongrel dogs). Cerebral Blood Flow and Metabolism. Acta Neurologica Scandinavica supp. 72 volume 60 - Ed; Fumio Gotoh, Hajime Nagai Yoshiaki Tazaki (Japan).
7. Cahn J. : This symposium.
8. Bentue-Ferrer D., Van Den Driessche J., Reymann J.M., Allain H., Bagot H. : Effect of triethyltin chloride on the cerebral aminergic neurotransmitters and their metabolites.Exp. Ag. Res. In press.
9. Borzeix M.G., Weber S., Akimjak J.P., Dupont J.M. (1978) : Effects of papaverine, dihydroergotoxine, tanakan (Ginkgo biloba extract) and vincamine (base and subtained release) on the biochemical and neurological disturbances due to a triethyltin induced edema in the rat. VII th International Congress of Pharmacology, Reims Satellite Symposium Cerebrovascular Diseases.

10.Gabard B., Chatterjee S.S. (1980) : Cerebral edema induced by triethyltin in the rat: Effects on an extract of Ginkgo biloba. Naunyn-Schmiedeberg's Arch Pharmacol. 311 : R 68.
11.Karcher L., Zagermann P. and Kriegelstein J. (1984) : Effect of an extract of Ginkgko biloba on rat brain energy metabolism. Naunyn-Schmiedeberg Arch Pharmacol. 327: 31-35.
12.Auguet M., Hellegouarch A., Delaflotte S., Baranes J., De Feudis F.V., Clostre F., Braquet P. and Drieu K. (1984) : Effects of Ginkgo biloba extract on rabbit isolated blood vessels. In : eds. A. Bes, P. Braquet, R. Paoletti and B.K. Siesjö, pp. 347-354. Amsterdam, Elsevier.
13.De Feudis F. : This Symposium.
14.Braquet P., De Feudis F.V., Deby C. and Braquet M. (1984) : Involvement of oxygen free radicals in biologic and pathologic processes. In : eds. A. Bes, P. Braquet, R. Paoletti and B.K. Siesjö, pp.265-283. Amsterdam, Elsevier.
15.Etienne A., Chapelat M., Braquet M., Clostre F., Drieu K., De Feudis F.V. and Braquet P. (1984) : In vivo studies of free radical scavenging activity; relation to cerebral ischaemia. In : eds. A. Bes, P. Braquet, R. Paoletti and B.K. Siesjö, pp. 379-384. Amsterdam, Elsevier.
16.Chatterjee S.S., Gabard B. (1982) : Protection of doxorubicin toxicity of an extract of Ginkgo biloba. Naunym-Schmiedeberg's Arch Pharmacol 319 : R 15.

GINKGO BILOBA EXTRACT

Clinical Studies

Effects of Ginkgo Biloba Extract on Organic Cerebral Impairment. A. Agnoli, J. R. Rapin, V. Scapagnini, W. V. Weitbrecht. John Libbey

Chronic cerebral insufficiency and Ginkgo biloba extract

P. F. Michel

IPSEN, 30 rue Cambronne, 75015 Paris, France

ABSTRACT

This communication gives some of the latest results obtained by pharmacological and clinical studies of the effects of Ginkgo biloba extract on chronic cerebral insufficiency. First of all a presentation of this original tree, whose leaves give us the therapeutic extract. Several studies have shown that GBE has a pharmacodynamic dose-related effect on vigilance, ably demonstrated by quantitative EEG analysis; GBE significantly increases the speed of memory scanning without effect on performance, wich immediatly differentiates it from simple behavioural stimulants; GBE increases cerebral blood flow and enhances glucose consumption; lastly GBE in clinical trials has shown proof of a marked beneficial effect upon sociability, vigilance, mood, memory and intellectual efficiency.

KEYWORDS

Ginkgo biloba tree - Ginkgo biloba extract - chronic cerebral insufficiency - human pharmacological and clinical studies - Alzheimer type dementia.

INTRODUCTION

Before introducing some of the latest studies carried out with Ginkgo biloba extract on chronic cerebral insufficiency we shall present the tree whose leaves give us specific molecules of great therapeutic use.

Gingko biloba (réf.1,3,6,7) is a tall, large, strong, sturdy and extremely long living (2000-4000 years) beautiful tree with a very curious type of leaf divided into two lobes in the young specimen whence the species name "biloba".

Dean of all trees (150.000.000 years or more) it has an unique place among present plants : last and only member of a vanished flora it is at one and the same time a species, a genus, a family, an order, a class and for some a phyllum.

It presents an original and archaic reproduction pattern: dispersing not real seeds but "ovules"; actually the Ginkgo tree is placed between the "spore plants" and the evoluated plants, the "seed plants".

Like many ancient vegetals, the Ginkgo elaborates a rich range of unusual compounds, some of them specific.

Ginkgo biloba has a long history of medical use in eastern Asia but it is only since the last world war that western pharmacologists became interested in the Ginkgo leaves extract. The use of GBE in France and other countries for some years came as a logical follow up after pharmacologic and pharmacoclinic investigation showing the beneficial effects of the drug on the circulatory system as a whole, and on tissue nutrition.

PHARMACOCLINICAL STUDIES

To limit the discussion to cerebral indications only of GBE we shall first mention two recent pharmacoclinical studies :

- The aim of the first one, carried out in West Germany by Krauskopf,is to demonstrate by quantitative EEG analysis the pharmacodynamic effect of GBE (ref.4). 6 young healthy volunteers received an acute dose of GBE (120, 240 and 600 mg) or a matched placebo. Authors observed with GBE a dose-related progressive increase in the power of the dominant frequency, an increase in the absolute value of α and β_1 dominant frequencies in the frontal derivations. The effect began during the first hour after drug intake and lasted at least 4 1/2 hours. This study clearly indicates a dose-related effect of GBE on vigilance as it is appreciated by EEG.

- The second double blind cross-over study was carried out by Subhan and Hindmarch (ref.8) in order to evaluate the influence of a range of acute oral doses of GBE on cognitive functions and information processing. The authors adopted a healthy volunteer strategy and used a battery of psychometric tests and a subjective rating scale. Memory was assessed using the Sternberg Memory Test in which performance depends on retrieval from short term memory. Authors found that a single dose of 600 mg of GBE increased the speed of memory scanning (Fig. 1).

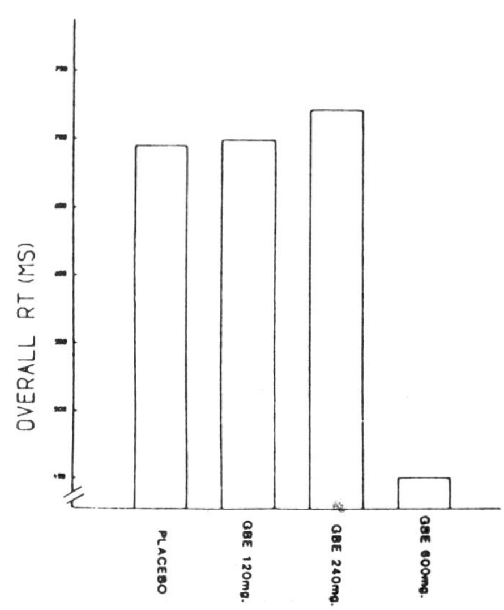

Fig. 1 : Overall reaction time on memory scanning test 1 h following treatment
. = significant difference from placebo ($p < 0.005$ or better).

Critical Flicker Fusion Test used as a measure of overall central nervous system activity and Choice Reaction Test used to assess sensorimotor performance did not show any increase of performance with GBE; this is in opposition to behavioural stimulants. This negative finding argues against GBE having a simple stimulant effect on performance and point to a more specific action on central cognifive processes. Subjective mood and sedation estimated by a 10 cm line analogue rating scale did not show significant change in healthy subjects. For the authors the collected data suggest a potentiel use for GBE in the treatment of patients suffering from senile or presenile dementia, in which impaired memory function is a prevailing characteristic.

The rational for the clinical use of GBE comes from the evidence that a proportion of patients with dementia have an impairment of neuronal metabolism. Consequently, a compound like GBE which increases blood flow and improves cortical metabolism should ameliorate the cognitive disorders which are characteristic of most type of dementia.

CLINICAL STUDIES

Overall analysis of clinical trials with GBE on chronic cerebral insufficiency first show that open trials, comparative trials, double blind studies gave sensibly similar results. Activity of GBE is particularly demonstrated on memory, intellectual efficiency, vigilance sociability, mood, headache and vertigo.

As an example Leroy and his collaborators (ref.5) have carried out a double blind comparative study with 50 inpatients suffering from moderate senile dementia. The assay was performed during two months versus an association of dihydroergotoxine-raubasine. Criteria of assessment were a rating scale of a Stockton type with 27 items and psychometric Benton test, Rey test, Trail marking test. After treatment only those patients who showed marked improvement were considered to have given positive results.

The control group (dihydroergotoxine-raubasine) gave some good results but they were significant for vigilance only. Patients treated by GBE show very significant improvement for sociability, vigilance and mood (Table I).

Table I. (Leroy and Coll.)

CLINICAL RESULTS (NORMALISATION OR BIG IMPROVEMENT)

	RYE E. + P.	G.B.E.
VIGILANCE	30.8% *	55.5% ***
SOCIABILITY	15.8%	52.6% **
MOOD	25 %	60 % ***

* $p \leqslant 0.05$ ** $p \leqslant 0.01$ *** $p \leqslant 0.001$

Similar to the preceding rating scale, psychometric tests studying intellectual efficiency show similar differences in favour of GBE (Table II).

Table II. (Leroy and Coll.)

	PSYCHOMETRY RYE E. + R.	G.B.E.
REY	41.0% *	63.0% ***
BENTON	9.0%	44.5% **
T.M.T.	22.6% *	52.0% ***

* $p \leqslant 0.05$ ** $p \leqslant 0.01$ *** $p \leqslant 0.001$

Another very demonstrative study is the one of Tea and Marc-Vergnes (ref.9) who investigated 19 patients suffering from transient or regressive cerebral ischemic accidents and their treatment by GBE during 2 months. The methods of assessment were measurement of the cerebral blood flow with technetium, measurement of cerebral glucose consumption, measurement of oxygen consumption and, clinically, a behavioural rating scale.

Table III shows that the mean cerebral transit time decreases significantly after treatment; this is in agreement with an increase of 16,6% of cerebral blood flow. Conversely, glucose consumption was markedly improved and the oxygen consumption about normalized. On the clinical side, behavioural rating scale shows a significant score increase after treatment.

Table III.

	Before treatment	After treatment	Δ	s.e.	p.
Behavioral scales - score...............	57.1	64.85	7.75	8.22	<0.01
Cerebral circulation time..............	14"48	12"08	2"40	2"40	<0.01
$\dot{V}O_2$ (ml/min./100 g)	2.76	3.37	0.61	1.08	<0.05
$\dot{M}$ glucose en mg/min./100 g....	2.76	3.85	1.10	2.28	<0.05
$\dot{M}$ lactate en mg/min./100 g....	- 0.025	- 0.030	- 0.005	0.038	N.S.

The last clinical study is the recent double blind study of Eckmann and Schlag of the University of Kiehl in Western Germany (ref.2). 50 patients suffering from chronic cerebral insufficiency received placebo or GBE (120 mg/j) during 1 month. Assessment was based on clinical symptomatology and neurological and mental signs. The overall results were especially good and very significant in the GBE group when compared to placebo (Fig. 2).

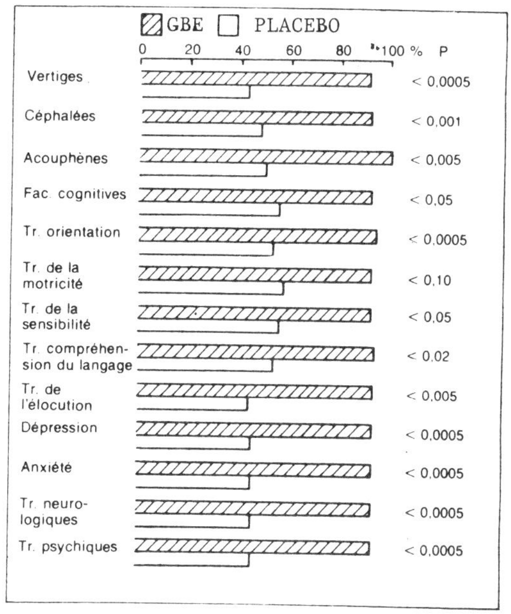

Fig.2: Disappearance of the symptom (%).

It is also very interesting to consider the rate of improvement . There is a dramatic and highly significant difference between the effect of placebo and GBE;

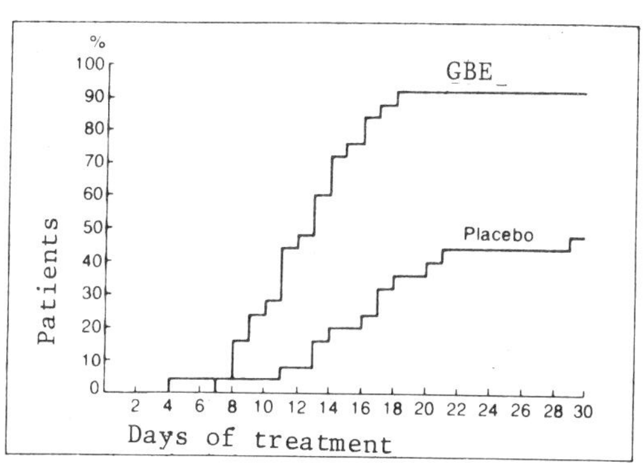

Fig. 3 : Speed of relief (p< 0.01)

CONCLUSION

At the present, we continue with pharmacological and clinical investigations aiming at a better comprehension of the mechanisms of GBE.
This is not easy; of this natural extract we know the exact composition of the three quarters only. Several fractions have been separated; they are active on several different processes but much less than the extract as a whole. Extract that, after many years, we have succeded in developing.

REFERENCES

1. Arnold C.A. (1947) : An introduction to paleobotany (Mc-Graw-Hill, New-York 1947) p. 227
2. Eckmann F., Schlag H. (1982) : kontrollierte Doppelbliendstudie zum Wirkungsnachweis von Tebonin forte bei patienten mit zerebrovaskulärer insuffizienz. Fortschr. Med. 31-32, 1474-1478.
3. Franklin A.H. (1959) : Virginia J. Sci. 10.131
4. Krauskopf R., Guinot P., Hauser B. (1984) : Long term on line EEG analyses demonstrating the pharmacodynamic effect of a defined Ginkgo biloba extract (in press).
5. Leroy M., Salaün P., Chovelon R., Bouilloux E., (1978) : Approche clinique et psychométrique en gériatrie. Méthodes d'études et choix d'une thérapeuthique. Vie Médicale 28, 2513-2519.
6. Lutz H.J. (1931) : Amer. Forests 37, 475.
7. Major R.T., (1967) : The Ginkgo, the most ancient living tree. In Science, vol. 157, 1270-1273
8. Subhan Z., Hindmarch I., (1984) : The psychopharmacological effects of Ginkgo biloba extract in normal healthy volunteers. Int. J. Clin. Pharm. Res. IV (2) 89-93.
9. Tea S., Celsis P., Clanet M., Marc-Vergnes J.P., (1979) : Effets clinique, hémodynamique et métabolique de l'extrait de Ginkgo biloba en pathologie vasculaire cérébrale. Gazette Médicale de France tome 86 n° 35 4149-4152.

Effects of Ginkgo Biloba Extracts on Organic Cerebral Impairment. A. Agnoli, J. R. Rapin, V. Scapagnini, W. V. Weitbrecht. John Libbey Eurotext Ltd © 1985

Long-term drug administration effects of Ginkgo biloba on the performance of healthy subjects exposed to hypoxia

K. Schaffler and P. Reeh

Laboratory of Electrophysiology, Munich, FRG

INTRODUCTION

Extr. Ginkgo bilobae* - respectively the inherent mixture of Ginkgo-flavonglykosides - is well known for a long time in the therapy of cerebral and peripheral perfusion problems. Positive drug effects seem to be due to multifactorial mechanisms on vascular, rheological and metabolic phenomena of the body, as demonstrated several times in pharmacological and clinical studies(3).

E.g. experimental approaches in animals have shown positive influence of the extract on brain edema (2), brain electrolyte balance (2), cerebral hypoxia tolerance (1), cerebral perfusion (4) and metabolic turnover, as well as on rheological problems (1).

But to verify such effects in humans -such robust methods cannot be introduced for drug testing procedures in the experimental or clinical routine. Furthermore well-known difficulties in setting-up a statistically homogenous group of patients with symptoms of "cerebral insufficiency" for demonstration of "encephalotropic" effects led to a model of partial simulation of pathophysiological conditions in the CNS - by generation of artificial hypoxia in healthy subjects.

The hypoxia administration model for testing antihypoxidotic, respectively encephalotropic effects in healthy volunteers has already been published elsewhere (10,11).

Another interesting approach in that hypoxia model was the introduction of the so - called Oculodynamic Test (ODT). It s principle - using simultaneous measurements of oculomotor and complex choice reaction system - has been published, too (12,13) and has already been used in several drug testing regimen. The basic idea of the computer-assisted ODT was the use of an extreme objective and sensitive test in psychopharmacological and encephalotropic drug research, because of its independence of learning effects and motivation (5,6,.14,15)

The oculomotor system, especially the saccadic eye movements, offers a lot of advantages for pharmacological studies in man, such as:

- an access to a "part of the brain" (the eye) by a "simple", external experimental approach
- an access to essential polysynaptic reflex mechanisms
- an access to complex afferent and efferent information processing of the CNS
- an access to retinal functions (retina contains glia cells!)
- an access to an organ submitted to intracerebral blood supply (a.carotis int., a. ophthalmica)

Since oxygen consumption is impaired in cerebrovascular pathology, it is important to know, whether a cerebroactive drug has an essential effect on cerebral oxygen turnover. Therefore the above-mentioned model was introduced - with its complexity - in the described Ginkgo biloba-study in healthy volunteers, too.

METHODS AND EXPERIMENTAL DESIGN

The study was performed on 9 healthy, male volunteers (mean age 26,0 years, mean body weight 72,6 kg) and was designed as a placebo-controlled, single-blind, crossover trial, comparing liquid Gingko biloba preparation with placebo, each administered for two weeks with a seven days´ wash-out period in between. The dosage of liquid Ginkgo biloba was 4 ml/day given in two divided doses (1 ml liquidum contains extr. Ginkgo bilobae e fol. sicc. 40 mg = 9 mg Ginkgoflavonglykosides) at a fixed time per day (8 a.m. and 3 p.m.).

Volunteers were assessed on day 14 of drug administration at the same day-time, to take account of the circadian characteristics of relevant physiological parameters. Subjects´ compliances were improved by taking "pseudourine-samples" to ensure regular intake.

Every assessment day consisted of 3 ODT-sessions - each lasting for 20 minutes - with a prevalue under room-air-conditions, another prevalue under hypoxic-conditions (10,5 % oxygen, 89,5 % nitrogen, inspiratory) and a post drug administration value 2 hours after a new acute intake of 2 ml Ginkgo biloba in the steady state.

The subjects were breathing a calibrated mixture of 10,5 % oxygen and 89,5 % nitrogen in a semiopen system via a mask with a buffer-bag in between, to enable volunteers to respirate pressure-free (see Fig. 1).

Under the hypoxic exposure subjects were asked to perform for 20 minutes the computer-assisted Oculodynamic Test (ODT), see Fig. 2. The ODT is an objective and very sensitive test, because saccadic eye movements are involuntary and independent of learning effects and motivation. The Oculodynamic Test examines the oculomotor and general information processing of the CNS with its perceptive, apperceptive and cognitive components (7,9).

With the head of the subjects fixed, light signals in well-defined locations of the horizontal visual axis were presented via a tube-lens-system and eye movements were recorded by electrooculography.

Light signals were presented randomly in type and in position. A choice reaction keyboard unit enabled the evaluation of complex choice-reaction-time and number of correct answers (see Fig. 2). A single test-session consists of a 20-minutes working period, with presentation of about 920 signals, causing 920 saccadic eye movements and same amount of choice reactions. Analog data and signals were computed in on-line-mode by a mini-computer (VARIAN V 73) after automatic calibration and artefact rejection.

The following data were obtained:

PARAMETERS OF ELECTROOCULOGRAPHY

1. Mean latency time (the time, which elapses between the appearance of the light signal in the periphery of the retina and the beginning of the eye movement for fixation of the target)

2. Mean angular velocity of saccadic eye movements

3. Accuracy of eye movement

OPERATIONAL PARAMETERS

1. Mean choice reaction time (answering the appearance of specific light signals on the keyboard)

2. Mean number of correctly identified signals

SIMPLE CARDIORESPIRATORY PARAMETERS

1. Mean pulse rate

2. Mean pulse arrhythmia

3. Mean respiratory rate

4. Mean respiratory arrhythmia

The findings of the hypoxia study were statistically evaluated by means of an adequate descriptive and analytical approach (2-way analysis of variance) with a consecutive test (SCHEFFE) in the event of significance.

RESULTS

ELECTROOCULOGRAPHICAL PARAMETERS

The polysynaptic retino-muscular **reflex-time ("latency")** of saccadic eye movements was impaired by hypoxia. Gingko biloba was able to improve this essential parameter highly significant (1%-level) and to reduce the prolonged reflex time under hypoxic conditions (see table 1).

The **angular velocity** of saccades, which is more dependent on pure muscular effects (mm. recti) was not significantly affected by hypoxia or drug administration.

Principle of the oculodynamic test (ODT)

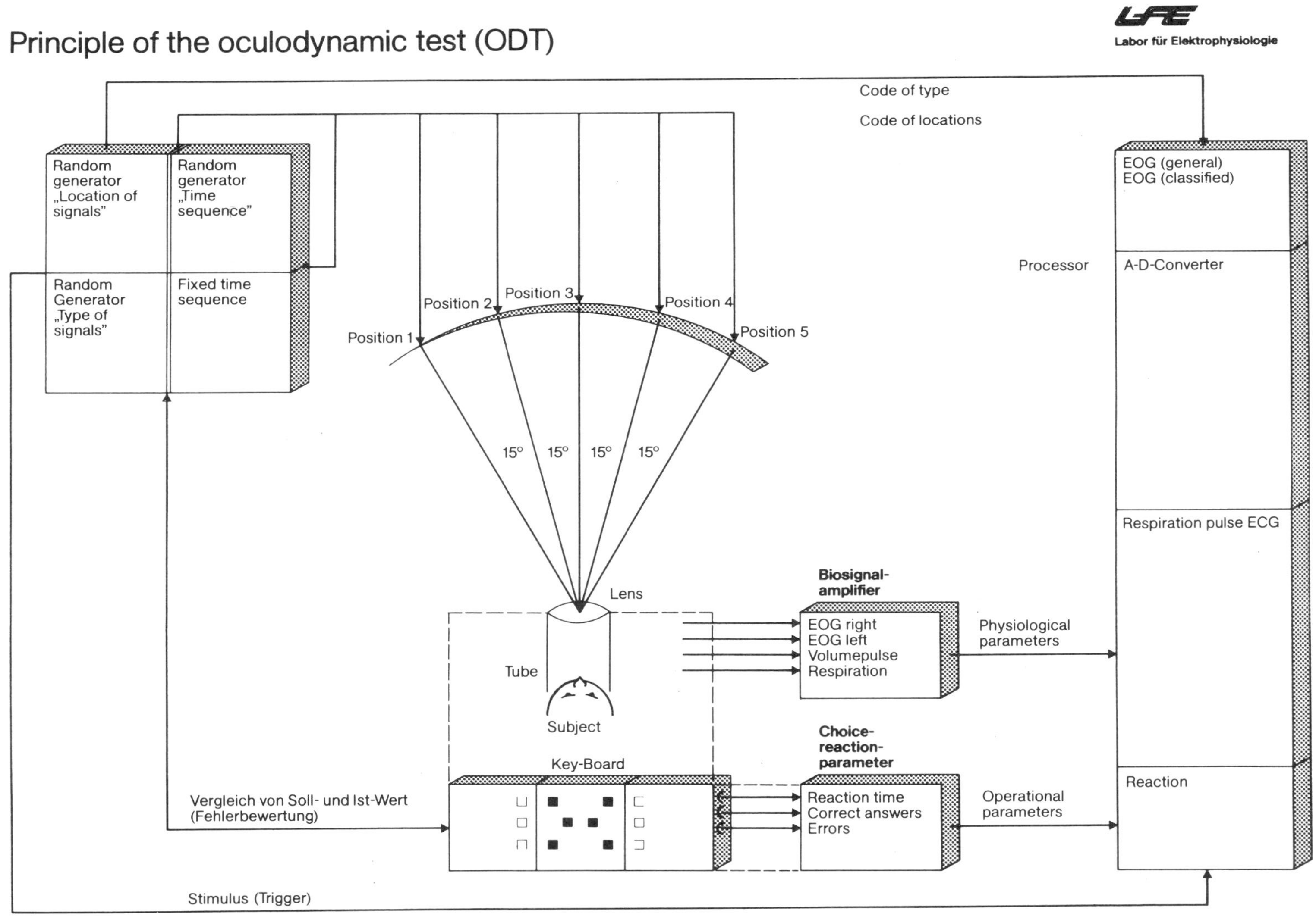

The prolonged fixation time was non-significantly reduced by Gingko biloba administration.

table 1: Results of Electrooculography (2h post medication values)

Latency (msec)	Placebo	141,2	(± 16,0)	**
	Verum	136,1	(± 11,5)	
Angular velocity (°/sec)	Placebo	568,7	(± 47,9)	n.s.
	Verum	564,1	(± 29,7)	
Fixation time (msec)	Placebo	71,6	(± 5,9)	n.s.
	Verum	70,1	(± 3,5)	

** : $p < 0,01$

OPERATIONAL PARAMETERS

The complex choice reaction time was non-significantly reduced by the verum preparation, whereas the total number of correctly identified light signals was not affected by the drug (see table 2)

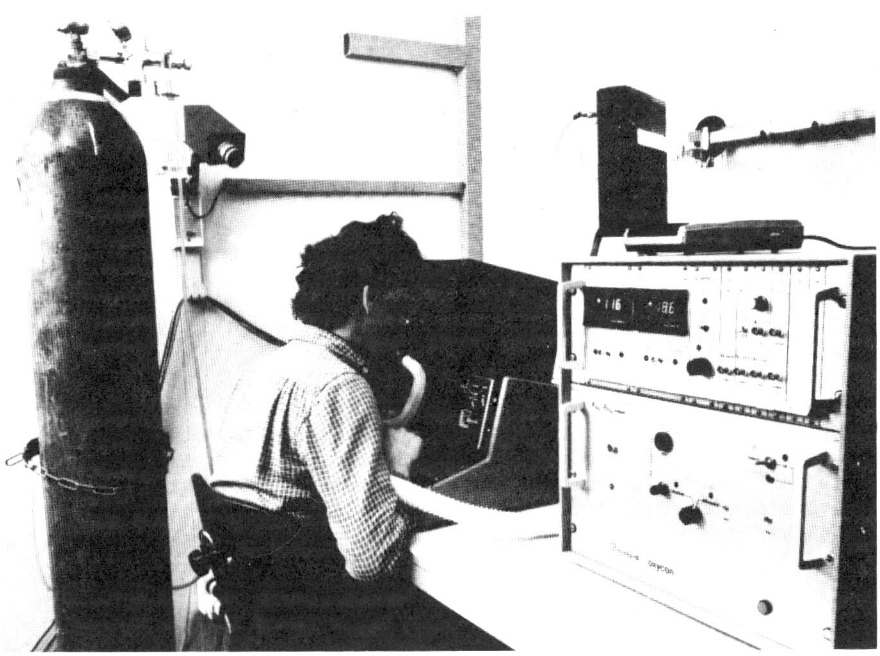

table 2: Results of operational parameters

Parameter	Group	Mean	(± SD)	
Choice react.time (msec)	Placebo	784,5	(± 120,6)	n.s.
	Verum	774,1	(± 103,1)	
Corr. identified signals (1/min)	Placebo	16,1	(± 6,2)	n.s.
	Verum	16,0	(± 6,2)	

SIMPLE CARDIORESPIRATORY PARAMETERS

All essential parameters, except the respiratory rate (5%-level), were not affected by Gingko biloba (see table 3).

table 3: Results of cardiorespiratory parameters

Parameter	Group	Mean	(± SD)	
Pulse rate (1/min)	Placebo	92,3	(± 11,1)	n.s.
	Verum	95,2	(± 7,9)	
Pulse arrhythmia (msec)	Placebo	19,6	(± 4,1)	n.s.
	Verum	18,3	(± 3,4)	
Respiratory rate (1/min)	Placebo	17,9	(± 4,2)	*
	Verum	16,6	(± 1,7)	
Respiratory arrhythmia (msec)	Placebo	230,2	(± 28,9)	n.s.
	Verum	243,1	(± 16,2)	

* : $p < 0,05$

All parameters of the computer-assisted Oculodynamic Test (ODT) revealed a smaller standard deviation under Ginkgo biloba application than under Placebo.

DISCUSSION

Due to the complex route of mechanisms of Ginkgo biloba on vascular, rheological and metabolic parameters, it is very difficult to give an exact explanation for drug effects on specific structures of the CNS. To elucidate these problems further, another study with electroretinography under hypoxia was done with Ginkgo biloba (looking at sodium/potassium pump and glia cells). which will be published in the near future.

Otherwise it is possible to describe a hypoxiaprotective mechanism of the drug on performance in healthy subjects after long-term application, which could mostly be demonstrated in such essential parameters like the polysynaptic retino-muscular reflex-time and the complex choice reaction time, which are part of basic principles of CNS-information processing and could be partly restored.

Another hint of improvement of oxygen consumption under drug application is the specific response of the respiration center after Ginkgo biloba administration, demonstrated by a reduced respiratory rate under hypoxia.

ABSTRACT

Long-term drug administration effects of Ginkgo biloba on the performance of healthy subjects-exposed to hypoxia

SCHAFFLER K., P. Reeh
(Laboratory of Electrophysiology, Munich, FRG)

The effects of liquid Gingko biloba (Ginkgoflavonglycosides) on hypoxia resistance of healthy young volunteers were measured by on-line monitoring of oculomotor and choice reaction phenomena. This model was developped for partial simulation of pathophysiological conditions to check "cerebral insufficiency improvers" in a homogenous group of healthy subjects.

The study was carried out in 9 healthy, male volunteers (mean age: 26,0, mean BW 72,6 kg) under single-blind conditions. Each subject received multiple doses of liquid Ginkgo biloba (2 ml, b. i. d. for 2 weeks) and placebo, each separated by a 7-days washout period.

Hypoxiaprotective (antihypoxidotic) effects of Ginkgoflavonglykosides could mainly be demonstrated by improvement of essential polysynaptic reflex mechanisms ("latency" of saccadic eye movements, complex choice reaction time) and improvement of respiratory response under hypoxic conditions versus placebo.

These results demonstrated that hypoxia seems to be an adequate model testing cerebral insufficiency improvers in healthy volunteers and supports results of animal experiments.

key words:

Ginkgoflavonglykosides, hypoxia, oculomotoric system, saccades, choice-reaction system, on-line monitoring

REFERENCES

1. CHATTERJEE, S. S. u. B. GABARD
Pharmakologisches Dossier
Dr. Willmar Schwabe, Karlsruhe (1980)

2. GABARD, B. u. S.S. CHATTERJEE
Naumyn-Schmiedebergs Archiv of
Pharmacology Suppl. Vol. 311, R 68 (1980)

3. HEISS, W.-D. u. K. ZEILER
Pharmakotherapie 1, 137-144 (1978)

4. ILIFF, L. D. u. L. M. AUER
J. Neurosurg, Sci. (1982)

5. KORNHUBER, H. H.
Blickmotorik
in Gauer/Kramer/Jung: Physiologie des Menschen
Bd. 6, S. 380-81, S. 391-92, Urban & Schwarzenberg Verl., (1978)

6. MACKENSEN, G.
Die Geschwindigkeit horizontaler Blickbewegungen
v. Graefes Archiv f. Ophthalmologie, 47:160 (1958)

7. MEYER-DELIUS, H., K. SCHAFFLER
Stress and Strain bei visueller Perzeption
Z. Arb. Wiss., 1:33 (1975)

8. PONCIN LAFITTE, M.le, J. RAPIN, J. R. RAPIN
Arch. int. Pharmakodyn. 243,
236 - 244 (1980).

9. SCHAFFLER, K., P. LINCKE, H. MAYER-DELIUS
Der Einfluß von Aufmerksamkeitsanforderungen auf Augeneinstellbewegungen, visuelle Perzeption und Rhythmik der Kreislaufreaktion
Ärztl. Praxis, 70:2679 (1975)

10. SCHAFFLER, K., E. HÖRMANN, H. ARNOLD
A pilot-study to evaluate the effect of acute and long-term administration of Cyclandelate on the vigilance of subjects submitted to hypoxic conditions
Drug Res., 31 (I), 6, 1032-1035 (1981)

11. SCHAFFLER, K., H. ARNOLD, E. HÖRMANN
A psychophysiological model to demonstrate CNS-protective mechanisms of a new 2-pyrrolidinone derivative (Ro 13-5057) in healthy subjects by on-line processing of the Oculodynamic Test (ODT) under hypoxic conditions
Abstract, World Congr. of Biol. Psychiatry, Stockholm, 1981

12. SCHAFFLER, K.,
The action of Clobazam and Diazepam on computer-assisted tests of muscle activity: Oculomotor effects
Drug Dev. Res., S1:39-46 (1982)

13. SCHAFFLER, K., H. ARNOLD, E. HÖRMANN
Vigilanzverhalten bei Interaktion des serotonin-selektiven Antidepressivums Zimelidin mit parenteral verabreichtem Alkohol
Drug Res. 32 (II), 8, 845-852 (1982)

14. VOSSIUS, G.,
Das System der Augenbewegungen
Z. Biol., 27:112 (1960/61)

15. WESTHEIMER, G.,
Eye movement responses to a horizontally moving visual stimulus
Arch. Ophthalmol., 52:932 (1954)

Effects of Ginkgo Biloba Extract on Organic Cerebral Impairment. A. Agnoli, J. R. Rapin, V. Scapagnini, W. V. Weitbrecht. John Libbey Eurotext Ltd © 1985

Clinical and psychometric evaluation of Ginkgo biloba extract in chronic cerebro-vascular diseases

A. Arrigo and S. Cattaneo

Universita Degli Studi di Pavia, Centro di Neurofisiologia Clinica

The aim of this work was to evaluate the efficacy of Ginkgo Biloba extract in patients suffering from chronic cerebro-vascular insufficiency.

EVALUATION	PARAMETERS
CLINICAL	Asthenia Headache Tinnitus Insomnia
PSYCHOLOGICAL	Anxiety (state & trait) Ability to concentrate Memory
PSYCHOMETRICAL	Logico-associative abilities Visuospatial abilities

Fig. 1

Evaluation parameters were clinical, clinico-psychological and psychometric.

The research was carried out as a double blind, placebo controlled, crossover trial, on a group of 80 patients, 29 females and 51 males, ranging in age from 40 to 80 years old.

Criteria for admission included patients suffering from vertigo, headache, tinnitus, loss of memory and of learning ability assuredly attributable to chronic cerebro-vascular insufficiency.

Patients were excluded from the trial if affected by serious neurological or mental deficits rendering them unable to co-operate, by clear-cut psychiatric disease, by chronic diseases liable to impair drug absorption, metabolism or catabolism, or by high risk. acute diseases (Such as recent miocardial infarction, etc.).

During the whole course of the trial, the use of other drugs, and particularly vasodilators, anti-Parkinsonians, anti-histaminic and psychoactive drugs was discontinued.

QUESTIONAIRE	S.T.A.I.	REY MT. WR.	WECHSLER
Asthenia	Tension	Attention	Learning
Insomnia	Anxiety	Memory	Mental deterioration
Headache	Depression	V.spatial & associtaive abilities	
Tinnitus	Emotional instability		
Visus impairment			

Fig. 2

Rating scales used to quantify baseline condition and patients' response to treatment both after placebo and after G.B. were respectively:

- a self rating questionaire for the clinical variables
- S.T.A.I. (State Trait Anxiety Inventory)
- Wechsler scale, memorizing tables, Rey shape type B and word recognition tables as regarding anxiety and psychometric parameters.

The research was scheduled as follows:

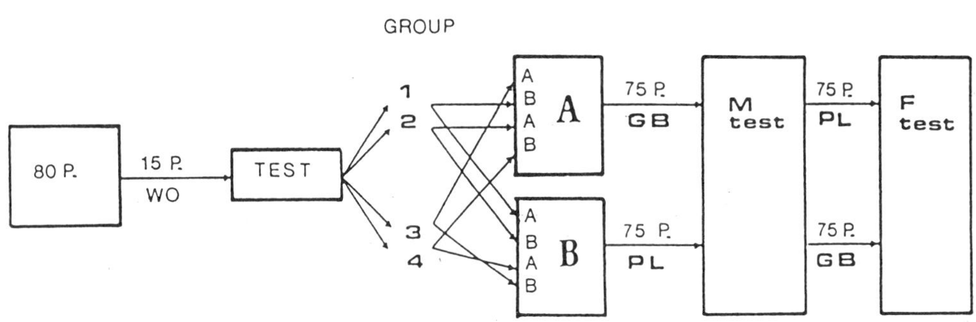

Fig. 3

After a pharmacological wash-out lasting two weeks, the eighty subjects underwent a series of tests and were then divided in four homogeneous groups for test results.

Each of these groups was finally randomly divided in two groups of ten, called respectively A and B.

Subjects from groups A were first given G.B. for 75 days, then retest, two more weeks' wash-out, finally placebo for another 75 days and a last retest. Groups B, on the other hand, were given first placebo and then G.B. Placebo and G.B. were identical in appearance and were both given per os, 50 drops thrice daily.

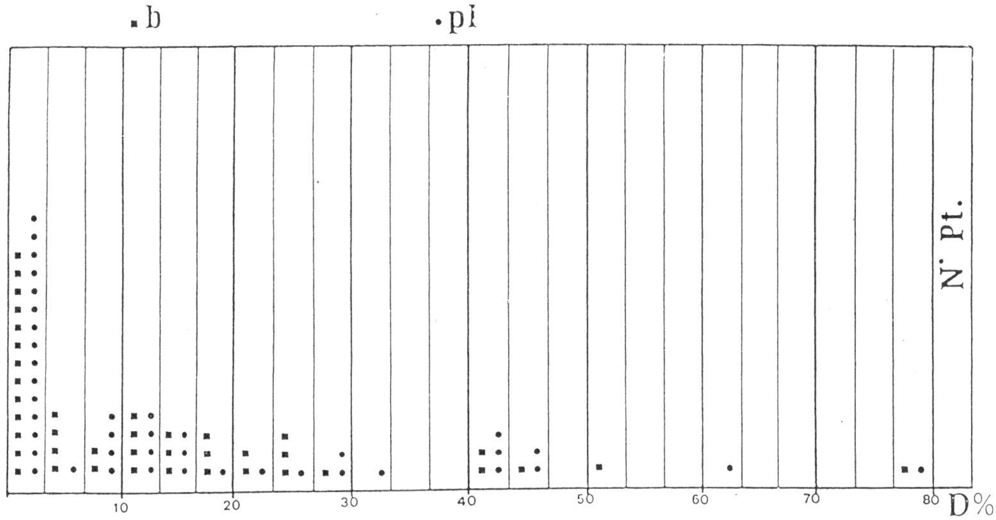

Fig. 4

This slide depicts Wechsler scale results obtained in the group of patients tested with placebo as compared with basal scores.
On the X axis = deterioration percentage
On the Y axis = number of patients

As can be seen, only a few patients in lower D% score group have improved, possibly thanks to a phenomenon of acquaintance to the test. On the contrary, there is a decided trend towards worsening in graver cases.

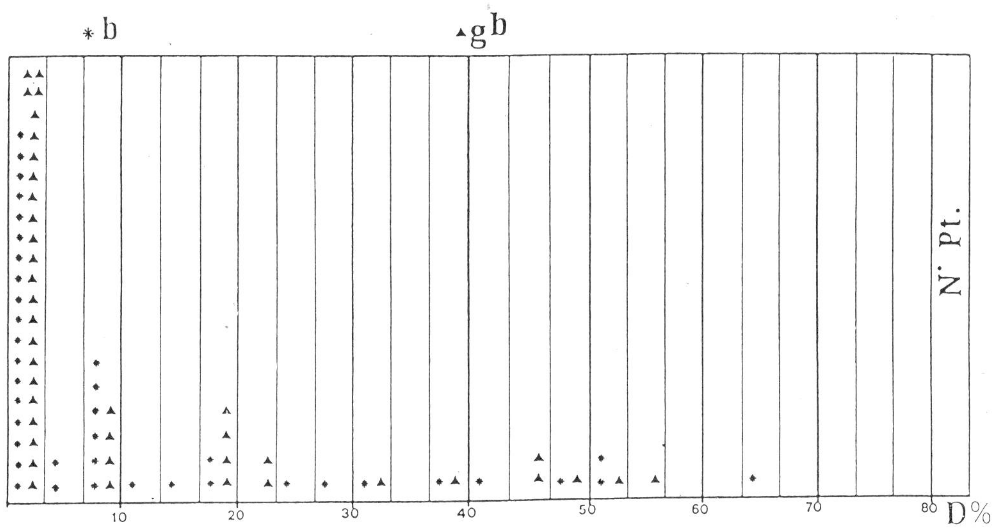

Fig. 5

These are the patients who were treated with G.B.; again basal and post treatment D% scores were compared.

Here we see both a marked improvement (in fact, 23 patients, versus 17 basal reach lowest D% class), and a decided shift to the left even in worse score groups.

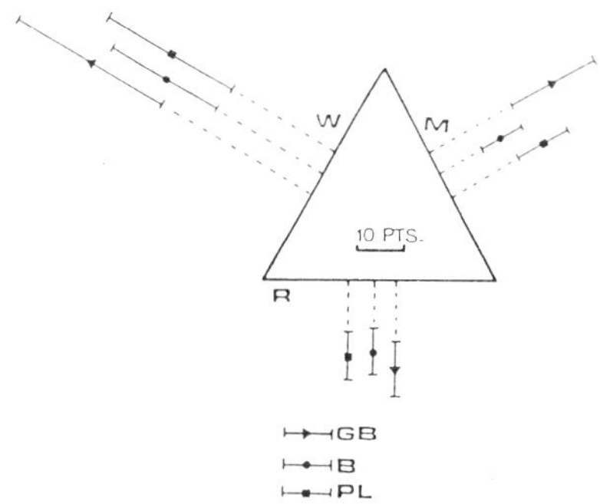

Fig.6

Among the psychometric variables considered, there were also memory and visuo-spatial abilities, which were investigated with the aid of Rey shape type B, memorizing tables (in which twelve different items were depicted in adjacent squares: patients could look for twenty seconds, then had to reproduce them in correct spatial order and as similar as possible to the originals) and word recognition tables (that is, ten couples of words, some identical, some very slightly different. Patients had to identify "false" couples in as little time as possible).

<u>Graph</u> : 0.5 cm every 10 points scored
Middle marks (dot) = mean of patients' scores with standard deviation basal test. Anticlockwise, marked with a triangle = mean scores and S.D. after placebo showing placebo improved scores in a not statistically remarkable way. On the contrary, G.B. produces a very significant increase in points scored in each of the three tests (P less than 0.05).

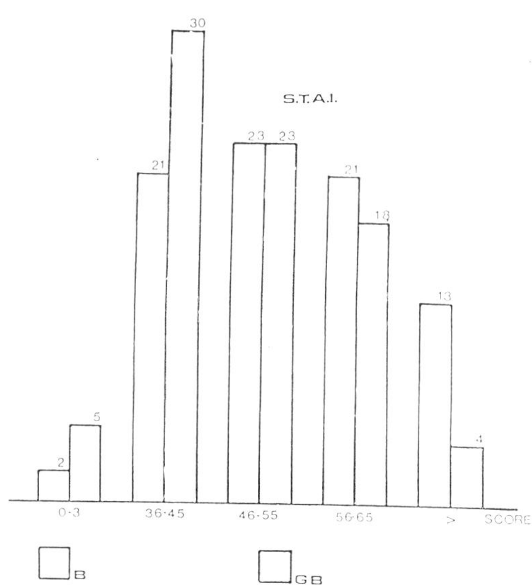

Fig.7

Anxiety was evaluated by means of the S.T.A.I., and results were as follows: Anxiety stated was greatly reduced by G.B. (X axis = S.T.A.I. score; Y axis = number of patients).

As can be seen, there is a general shift to lesser involvement scores. Against that, we found that anxiety trait could not be influenced by drug administration, appearing thus to confirm the opinion of those who declare it to be an intrinsic personality trait, not modifiable by therapy. Now we come to the variables investigated with the aid of the self-rating questionaire.

This was repeated eight times during the course of the trial, that is four before, during and after placebo, and four before, during and after G.B.

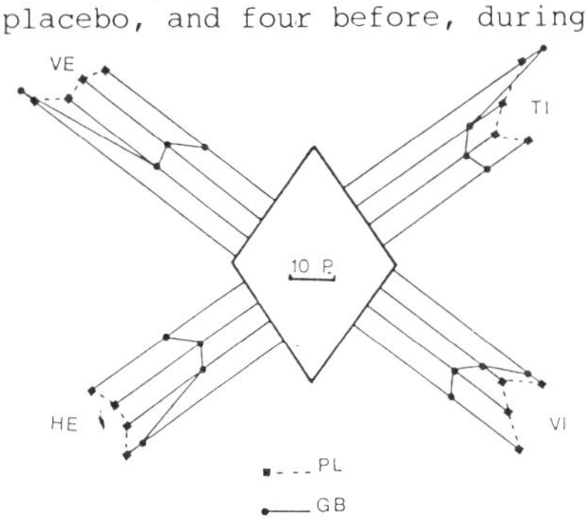

Fig.8

Here 0.5 cm. represents ten patients. Dot = G.B. trial. Square = placebo

Tebonin appeared to be significantly effective on vertigo, headache, tinnitus, visual impairment. Placebo effect, through present, is not in the least as significant, and tends in most cases to abate towards the end of the trial.

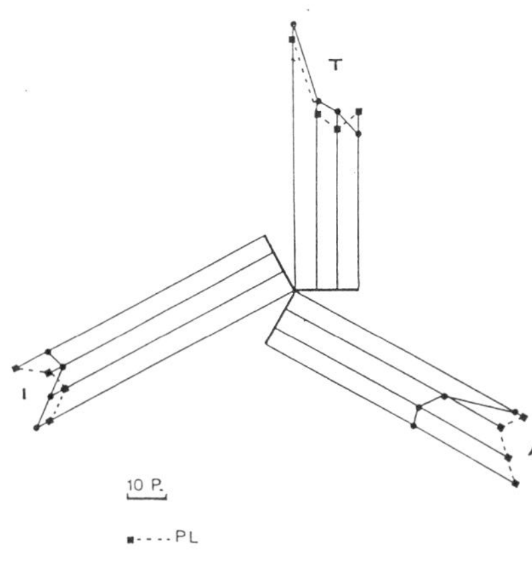

Fig.9

G.B. also markedly diminished the number of patients suffering from asthenia. On the other hand, both G.B. and placebo were ineffectual against insomnia. Tension, on the contrary, was extremely sensible to treatment both with placebo and G.B., but probably depending more on the newly-built strict therapeutic relationship than on any drug (in fact, it was high at the beginning, after suspension of former therapies and the decreased remarkably and quite comparably both after placebo and G.B.).

Effects of Ginkgo Biloba Extract on Organic Cerebral Impairment. A. Agnoli, J. R. Rapin, V. Scapagnini, W. V. Weitbrecht. John Libbey Eurotext Ltd © 1985

Doubleblind and comparative (Ginkgo biloba versus placebo) therapeutic study in geriatric patients with primary degenerative dementia - a preliminary evaluation

W. V. Weitbrecht and W. Jansen

Neurologische Universitaetsklinik Erlangen and Fritz-Hintermayer-Altenkrankenhaus, Nuremberg, Germany

ABSTRACT

Ginkgo biloba extract (GBE) in therapy of slight and moderate primary degenerative dementia has been assessed using a double blind randomized study versus placebo. An homogeneous population of 40 patients (age : 60-80) has been treated by GBE (120 mg/day) or placebo for a period of three months. Treatment's evaluation was based on own patient assessment, clinical geriatric scales (Crichton, SCAG) and psychometric tests (Weshler) carried out before and after 4, 8 and 12 weeks of treatment.
Results show from the eighth week an improvement in GBE group which seems to be clinicaly relevant. Indeed, psychometric tests but also clinical scales, which juge the everyday outcome of patient, show statistically significant improvement.

KEYWORDS

Ginkgo biloba extract - Primary degenerative dementia - clinical study - double blind placebo - clinical geriatric scale - psychometric tests -

INTRODUCTION

Improvement of outcome of patients with primary degenerative or multi-infarct dementia by drugs is a well known problem. The great number of papers dealing whith this subject show the difficulties to demonstrate the action of different nootropica. Most of the drugs which are increasing peripheral blood flow do not change or even worsen cerebral blood flow (Herrschaft 1976). Some drugs change electroencephalographic parameters. But it is questionable wether these changes are relevant for outcome of geriatric patients. Clinical rating scales with an acceptable symptom profile of dementia in geriatric patients are therefore as essential in the evaluation of drug therapy as appropriate psychometric tests (Venn 1983). Many investigations are insufficient, since they examine an inhomogenous population of patients, unproper and too much parameters and too short a period of therapy (Weitbrecht 1983).

The following study investigates therapeutical efficacy of Ginkgo biloba extract in patients with primary degenerative dementia for a period of three months. Diagnosis was made by Hachinski scale (Hachinski and al. 1975,table 1).

Table I : Hachinski scale (ischaemic score)

Abrupt onset	2
Stepwise deterioration	1
Fluctuation	2
Nocturnal confusion	1
Relative preservation personality	1
Depression	1
Somatic complaints	1
Emotional lability	1
Hypertension	1
History stroke	2
Focal symptoms	2
Focal signs	2
Other signs of arteriosclerosis	1

METHODS

40 patients aged 71.7 ± 4.6 years (26 females, 14 males) were taken in this study on the following premises :

1. slight or moderate dementia
2. Hachinski scale score less than seven
3. aged between 60 and 80 years

Patients with other chronic diseases as e.g. alcoholism, drug addiction, severe diabetes and malignomas were excluded. Two patients dropped out before the study was finished :
no. 41 (R.M., female, 76 y's) underwent an operation of cholelithiasis in the third week and no. 59 (G.K., female, 70 y's) with an appendicitis.

The study was layed out doubleblind randomized :
1. 20 patients treated with placebo
2. 20 patients treated with Extr. Fol. Ginkg. bilob. sicc. puriss. (Ginkgo biloba), standardized on ginkoflavonglycosid, 120 mg/die.

Preceeding psychotropic therapy was discontinued 2 weeks before study (drugs containing nicotinate were taken by 51 patients, pentoxyphylline by 5, inosin by 9). The mean duration of dementia symptoms was 8.8 ± 2.7 years. The general state of health was scored by the physician (excellent, good, poor, bad) at the beginning and the end of therapy. Chi-square test showed, that the two groups did not differ with age, Hachinski-scale score, duration of symptoms, and state of health at the beginning of the study (table 2).

Table II : Age, sex, Hachinski scale-score, duration of symptoms, state of health of the patients.

	All patients	Placebo	Ginkgo biloba
Age	71.7 ± 4.6	71.5 ± 4.4	71.9 ± 5.8
n	40	20	20
male/female	14/26	7/13	7/13
Hachinski-scale score	4.5 ± 1.0	4.6 ± 1.0	4.5 ± 1.0
duration of symptoms (years)	8.8 ± 2.7	8.8 ± 3.0	8.9 ± 2.5
State of health (physician) before therapy	3.2 ± 0.6	3.1 ± 0.6	3.3 ± 0.5
after 12 weeks	2.7 ± 0.9	3.1 ± 0.8	2.4 ± 0.9
n (after 12 weeks)	38	18	20

To evaluate the effect of therapy the following scales and tests were used:

1. the patients scored their general feeling (excellent, good, poor, bad)
2. Crichton-geriatric-scale (Table 3)
3. Sandoz clinical assessment geriatric scale (Table 4)
4. figure-symbol-test (Wechsler)
5. number-repeat-test (Wechsler)
6. flicker-fusion-frequency
7. reaction time

1-5 were carried out at the beginning and after 4,8, and 12 weeks of treatment. 6 and 7 were executed only at the beginning and the end of the study. The total score only was exploited of the geriatric scales (2,3). Statistical analysis was performed by chi-square test and u-test, (Kruskal-Wallis).

Table III : Crichton geriatric scale

No	Items	Score
1	mobility	1 = not pathological
2	orientation	2 = very mild
3	communication	3 = mild
4	cooperation	4 = moderate
5	restlessness	5 = severe
6	dressing	
7	feeding	maximum score : 55
8	continence	
9	sleep	
10	mood objective	
11	mood subjective	

Table IV : Sandoz clinical assessment-geriatric scale

No	Items	Score
1	confusion	1 = not present
2	mental alertness	2 = very mild
3	impairment of recent memory	3 = mild
4	disorientation	4 = mild to moderate
5	mood depression	5 = moderate
6	emotional lability	6 = moderately severe
7	self care	7 = severe
8	anxiety	
9	motivation-initiative	
10	irritability	maximum score : 133
11	hostility	
12	bother some	
13	indifference to surroundings	
14	unsociability	
15	uncooperativness	
16	fatigue	
17	appetite (anorexia)	
18	dizziness	
19	overall impression of patient	

RESULTS

The state of health scored by the physician did not change in patients treated with placebo (score at the beginning 3.1 ± 0.6 at the end 3.1 ± 0.8, but changed significantly ($p = 0.0065$ u-test) in patients treated with GBE (Table 2). Just so the subjective score of their feeling and sleep did not change in patients treated with placebo, in contrast patients treated with GBE reported a significant ($p<0.0048$ u-test) improvement of feeling and sleep in the 8th and 12th week of therapy. Appetite did not change in all groups. Compared to preceeding therapy the physician judged that patients treated with placebo did not improve. In contrast patients treated with GBE got better significantly ($p = 0.0001$ u-test score : better, no difference, worse).

Flicker-frequency increased significantly in patients treated with GBE from the beginning to the end of therapy (Table 5). Patients treated with placebo did not improve. Corresponding reaction-time decreased in patients treated with verum but not with placebo (Table V).

Table V: Flicker-frequency and reaction-time before therapy and at the end of the study (* = significant difference).

	Placebo	Ginkgo biloba
Flicker-frequency before (hz)	38.9 ± 2.1	39.6 ± 2.3
After 12 weeks	38.5 ± 1.9	41.4 ± 2.4 *
Reaction-time before (sec)	1.85 ± 0.32	1.96 ± 0.3
After 12 weeks	1.83 ± 0.26	1.63 ± 0.28 *

Scores of figure-symbol-test and number-repeat-test increased significantly in patients treated with GBE ($p<0.005$) but not in patients treated with placebo (figures 1 + 2).

Fig. 1: Figure-symbol-test scores (P = Placebo, G = Ginkgo biloba, + significant change p< 0,005).

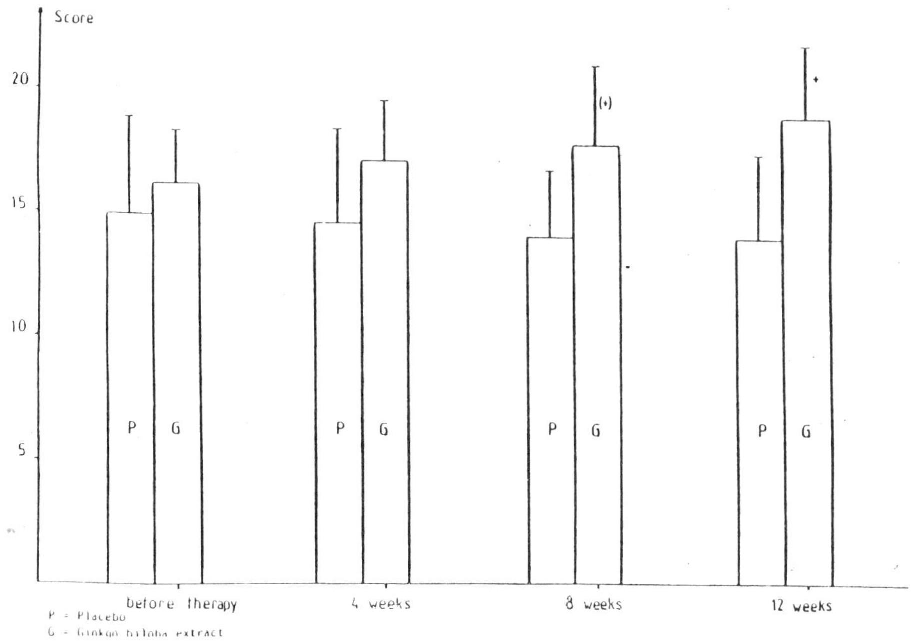

Fig. 2 : Number-repeat-test (P = placebo, G = Ginkgo biloba, + significant change p< 0,005).

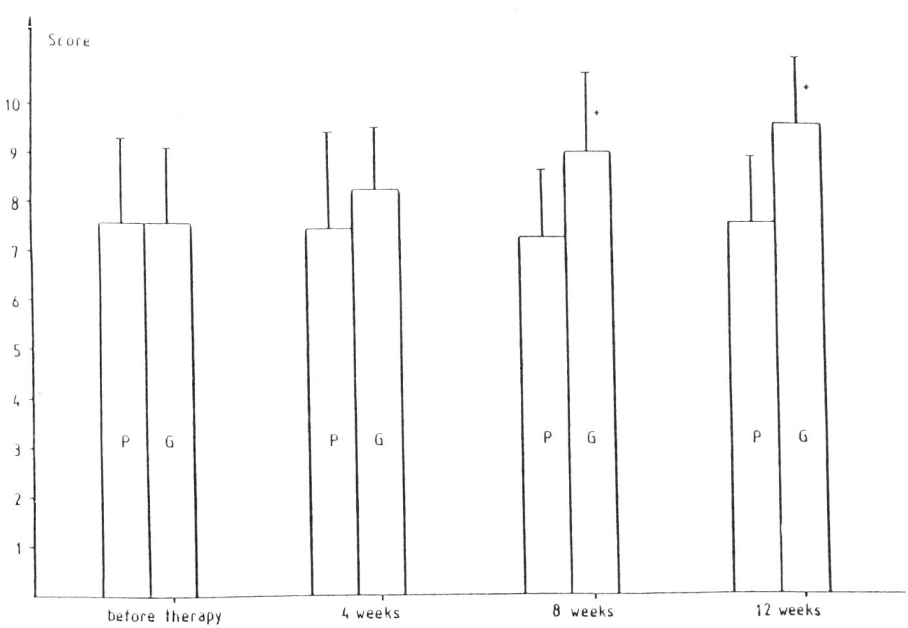

The total scores of the geriatric scales (Crichton scale, SCAG), decreased significantly in the 8th and 12th ($p<0.0005$ each) week in patients treated with GBE but not in patients treated with placebo (figures 3 + 4).

Fig. 3 : Crichton geriatric scale - total score (P = placebo, G = Ginkgo biloba).

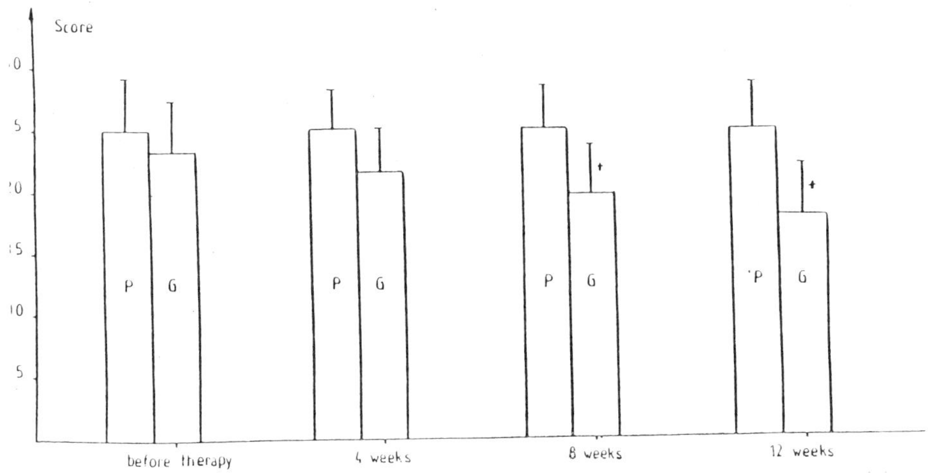

Fig. 4 : Sandoz clinical assessment-geriatric scale - total score.

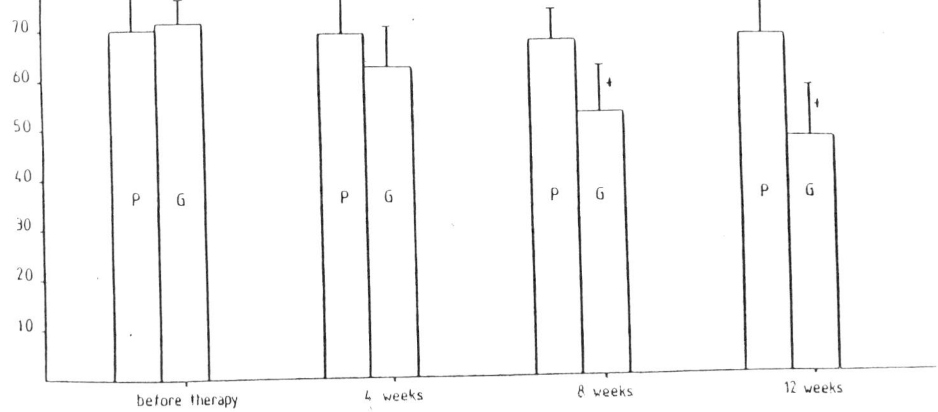

DISCUSSION

In studies on drug therapy of dementia of old patients it is important to investigate an homogeneous population. So it is possible to decide, whether the used drug is able to improve a certain disease e.g. multi-infarct dementia. Investigations on inhabitants of an old-age asylum (e.g. Biel and al. 1976, Kugler and al. 1978) are often unsatisfactory because of their lack of well defined diagnostic subgroups.The differential diagnosis between vascular and primary degenerative dementia is a common clinical problem. The ischaemic score of Hachinski and al. (1975) was a development of the use of clinical criteria which have been shown (score >7) to correlate with electroencephalographic and angiographic abnormalities, reduced cerebral blood flow (Harrison and al. 1979) and postmortem evidence of multiple infarction of demented subjects (Corsallis 1962, Slater and Roth 1969, Roth 1971). Score less than seven labeled primary degenerative dementia. So it is evident that the Hachinski-scale is a valid diagnostic tool for differential diagnosis of dementia as used in this investigation. The two subgroups in this doubleblind-comparison study diagnosed by the Hachinski-scale as patients with primary degenerative dementia proved to be not different (except sex).

The study shows that not only the scores of the psychometric test but also the scores of the clinical geriatric scales improve significantly after 8 and 12 weeks therapy with GBE but not with placebo. After 4 weeks therapy we found no really significant differences. So only an observation period of more than 4 weeks shows the efficacy of drug therapy.

The use of total scores of symptoms to evaluate efficacy of drugs as used in this study has been suggested by other authors too (Braconnier, Cole 1978).

CONCLUSION

The improvement by GBE therapy seems to be of clinical relevance, because not only psychometric tests but also the clinical geriatric scales which judge the everyday outcome of patients show significant improvement. The way of action of these drugs work is not really understood. An influence on cerebral metabolism is discussed.

Acknowlegement: The authors thank Mrs. Ulrike Weitbrecht for drawing the figures and typing the tables.

REFERENCES

1. Biel M.L., Seus R., Struppler A. (1976) : Medikamentoese Therapie des hirnorganischen Psychosyndroms im Alter. Med. Klin. 71, 2177-2184
2. Braconnier R.J., Cole J.O. (1978) : Total symptom index as a symptom independent parameter of drug efficacy in geriatric psychopharmacology. J. Gerontol. 33, 217-223.
3. Corsallis J.A.N. (1962) : Mental illness and the ageing brain. Oxford University Press, London.
4. Hachinski V.C., Iliff L.D., Zilhka E., Du Boulay G.H., Mc Allister V.L., Marshall J., Russel R.W.R., Symon L. (1975) : Cerebral blood flow in dementia. Arch. Neurol. 32,632-637.
5. Harrison M.J.G., Thomas D.J., Marshall J. (1979) : Multi-infarkt dementia. J. Neurol. Sci. 40,97-103
6. Herrschaft H. (1976) : Die Therapie der cerebralen Mangeldurchblutung. Nervenarzt 47,639-650
7. Kugler J., Oswald W.D., Herzfeld U., Seus R., Pingel J., Welzel D. (1978) : Langzeittherapie altersbedingter Insuffizienzerscheinungen des Gehirns. Dsch.Med.Wschr. 103,456-462.
8. Roth M. (1971) : Classification and aetiology in mental disorders of old age. Some recent developments. In: D.W.K. KAY and A.WALK (Eds.), Recent Developments in Psychogeriatrics. Headley Bros., Ashford, p.1.
10. Slater E., Roth M. (1969) : Clinical psychiatry. 3rd edition, Balliere, Tindall and Cassell, London.
11. Venn R.D. (1983) : The Sandoz clinical assessment-geriatric (SCAG) scale. Gerontology 29,185-198.
12. Weitbrecht W.-U. (1983) : Moeglichkeiten der Wirksamkeitskontrolle von "Nootropika". Terapiewoche 33,1343-1350.

Effects of Ginkgo Biloba Extract on Organic Cerebral Impairment. A. Agnoli, J. R. Rapin, V. Scapagnini, W. V. Weitbrecht. John Libbey

Effects of Ginkgo biloba extract on Xenon 133 CBF and I 123 iodo-amphetamine distribution in human brain acute infarction: a preliminary study

G. Rancurel[1], C. Raynaud[2], J. Cahn[4], E. Cabanis[5], K. Drieu[6], M. Allard[6] and N.A. Lassen[3].

[1]*Clinique Neurologique du Pr. A. BUGE (Hôp. Salpétrière), Paris*
[2]*Service Hospitalier Frédérik Joliot, 91406 Orsay, France*
[3]*Department of Clinical Physiology/Nuclear Medicine, Bispebjerg Hospital, 2400 Copenhague NV, Denmark*
[4]*Departement de Thérapeutique Expérimentale, SIR International 92120 Montrouge, France*
[5]*Hôpital des Quinze-Vingts, Paris*
[6]*IHB-IPSEN, Paris*

ABSTRACT

Ginkgo biloba extract (GBE) was administered to 6 patients with a unilateral MCA infarct during the first month which will be referred to as the acute period. Before and during this treatment rCBF and IMP distribution were studied with a high sensitivity SPECT system, TOMOMATIC 64 ; both were controlled later after 3 months in the chronic period. Infarct, cortical in four patients, was visible on CT scan with contrast enhancement. It showed a relative luxury perfusion, i.e. a higher rCBF in the acute period than in the chronic period, and a delayed IMP (5th hour) hyperfixation. In 3 of these patients GBE induced a decrease of rCBF in the relative luxury perfusion area, and in the rest of the brain ; it also induced a decrease in the delayed IMP hyperfixation. The rCBF and delayed hyperfixation decrease was present in the fourth patient but not significant. When GBE treatment was discontinued, rCBF and delayed IMP hyperfixation increased again. In 2 of these cases GBF was given from the very first day of the stroke with a maximal effect on delayed IMP hyperfixation ; both patients had a complete clinical recovery. Two other patients with a small and deep infarct were treated with GBE without any modification of rCBF and IMP kinetics in one and with an rCBF increase in the other. Significant changes can be obtained with GBE when given in infarcts during the acute period, but they were only observed in cases of superficial infarct.

KEYWORDS

rCBF Xenon 133, relative luxury perfusion, I-123-iodoamphetamine, IMP kinetics, IMP kinetics, IMP delayed hyperfixation, Ginkgo biloba, M.C.A. infarcts.

INTRODUCTION

Experimental studies have demonstrated that various therapeutic procedures are effective because they either minimize the ischemic process itself or increase brain's tolerance to ischemia (5). No general anti-ischemic therapeutic procedure exists, however Ginkgo biloba has shown an undoubted effect on experimental acute brain ischemia. In this preliminary work we studied the effects of Ginkgo biloba on the regional cerebral blood flow (rCBF) and the distribution of I 123 iodo amphetamine (IMP) in human acute infarcts.

MATERIAL AND METHOD

Six patients, 55 to 73 years old with unilateral completed infarct in the middle cerebral artery (MCA) territory due to embolism from atherosclerotic occlusive or stenotic disease of the internal carotid artery (ICA), were studied. A high sensitivity SPECT system, the TOMOMATIC 64, was used to measure the IMP brain distribution (1,3,10) with Xe 133 the rCBF, on the same SPECT slices (3,4). Xe 133 was administered by inhalation (3), IMP labelled with kits (ORIS) was injected intravenously at a dose of 6 to 8 mCi ; specific activity was 1 mg/mCi. IMP imaging was performed 10 min., 2 and 5 hours after injection (6). Ginkgo biloba extract (GBE) was administered per os, 340 mg per day. In the initial period, 0 to 30 days, which will be referred to as the acute period, rCBF and IMP distribution were studied with and without GBE administration ; they were measured again later after 3 months in the chronic period. In 2 patients GBE was administered from the very first day of stroke ; rCBE and IMP distribution were initially studied during GBE administration on the 16th and 11th day and then without GBE ; in the 4 others, rCBF and IMP were studied before GBE administration, on the 3rd, 13th and 19th day and then on the 4th and 5th day of GBE administration.

During each sequence of rCBF and IMP tests, the patients had CT scan without and with contrast enhancement, and slices were obtained at the same level as TOMOMATIC slices. Each patient had four-vessel X-ray angiogram.

MCA infarct was superficial in 4 patients, it was deep capsular or capsulo-lenticular in the two others. These 6 patients had hemiplegia with or without sensory defect or aphasia but did not show lethargy, stupor or any confusional state.

RESULTS

In four patients, infarct area rCBF was higher in acute period than in the chronic period, + 67 %, 323 %, 143 % and + 14 %, and was called "relative luxury perfusion". Only one of these patients had an effective luxury perfusion with an infarct area rCBF higher than the control area rCBF (Table I). They also had a delayed 5th hour IMP hyperfixation although they had a 10 minute IMP hypofixation, on the infarct area. Hyperfixation means that the infarct area fixation was higher than the contralateral area ; individual values were + 5 %, + 27 %, + 7 % and + 10 %.

When these 4 patients with relative luxury perfusion and IMP delayed hyperfixation were treated with GBE, rCBF and hyperfixation significantly decreased on the infarct area in three (Fig A and B) ; it decreased also on the fourth but not significantly. All four had a superficial MCA infarct and CT contrast enhancement ; in three GBE also led to a decrease in rCBF on the contralateral hemisphere.

In the 5th case with capsulo-lenticular infarct, rCBF did not decrease in infarct area compared to chronic period values and GBE did not induce any change in rCBF or IMP fixation on infarct area as well as on contralateral hemisphere.

PATIENTS	ACUTE PERIOD RESULTS: DATE	ACUTE: r C B F: Infarct area ml/mn/100g	ACUTE: r C B F: Contralateral area ml/mn/100g	ACUTE: r C B F: Infarct area decrease or increase in % of contralateral	ACUTE: IMP infarct area decrease or increase in % of contralateral area: 10'	ACUTE: IMP: 2 H	ACUTE: IMP: 3 H	CHRONIC PERIOD RESULTS: DATE	CHRONIC: r C B F: Infarct area ml/mn/100g	CHRONIC: r C B F: Contralateral area ml/mn/100g	CHRONIC: r C B F: Infarct area decrease or increase in % of contralateral	CHRONIC: IMP infarct area decrease or increase in % of contralateral area: 10'	CHRONIC: IMP: 2 H	CHRONIC: IMP: 3 H
JUI..	d 13	45	47	- 4	- 8	+ 3	+ 5	d 95	27	57	- 53	- 40	- 24	- 19
	d 19 +GBE	43	52	-17	-20	- 8	+ 8							
MEL..	d 19	55	55	0	- 9	+14	+27	d 180	13	45	- 71	- 45	- 54	- 48
	d 24 +GBE	29	36	-20	-13	+ 7	+20							
WEI..	d 11	51	47	+ 8	- 2	- 2	+ 7	d 210	21	40	- 48	- 38	- 39	- 31
	d 5 +GBE	43	41	+ 5	-12	-23	-22							
DEC..	d 26	33	49	-33	-20	- 7	+10	d 135	29	43	- 33	- 33	- 16	-. 9
	d 18 +GBE	28	36	-22	- 8	- 1	+ 5							
LEC..	d 6	36	37	- 3	- 3	- 2	0	d 180	38	43	- 12	- 7	- 6	- 3
	d 12 +GBE	35	36	- 3	- 2	- 1	- 1							
GIR..	d 3	34	44	-23	-13	-10	- 7	d 240	37	43	- 14	- 14	- 8	- 9
	d 8 +GBE	45	61	-26	-14	- 3	- 6							

TABLE I

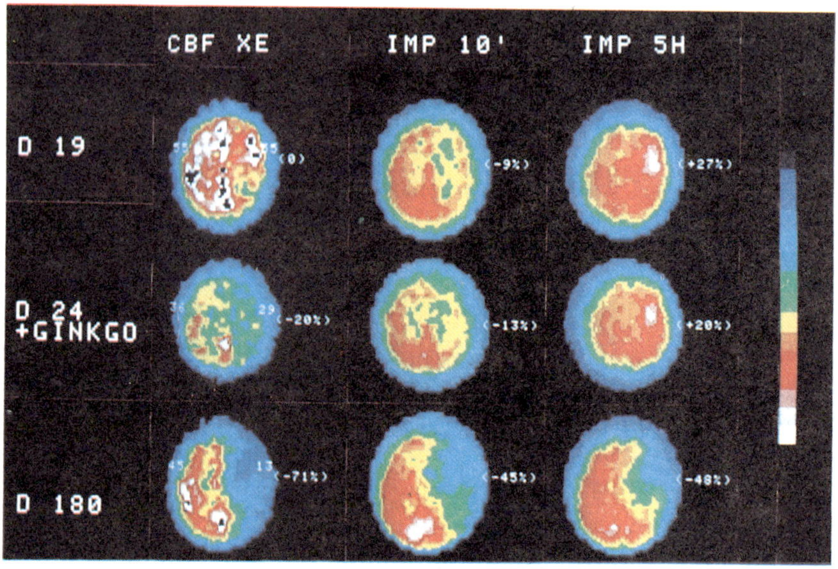

FIG A. CASE No 3

Patient MEL... with a right MCA Infarct. CBF measured with 133-Xe and IMP studies were repeated three times, at 19th day, 24th day and later at 180th day. Ginkgo Biloba extract was administered from 19th to 25th day. Regional, CBF values in ml are indicated on the relative luxury perfusion area and the symmetrical area. The decrease or increase of CBF and IMP expressed in % of contralateral value is inscribed in brackets.

FIG B CASE No 1

Patient JUI...with a left MCA infarct. CBF and IMP were repeated 3 times, 13th day, 19th day and 97th day, and Ginkgo Biloba extract was administered from 13th day to 20th day. CBF values and relative decrease of regional CBF and IMP fixation are indicated as in Fig. 1. The relative luxury perfusion observed on 13th day is attenuated during Ginkgo Biloba treatment.

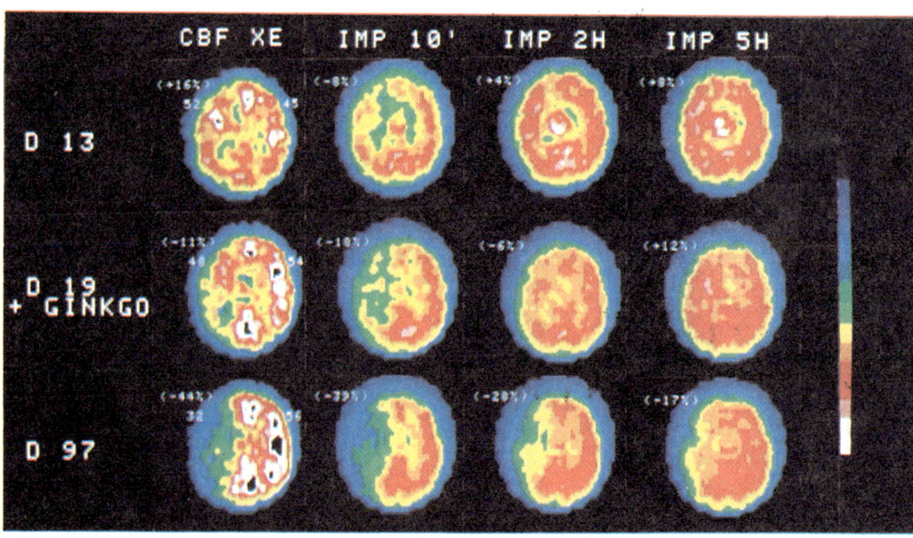

In the 6th patient with a capsular infarct, rCBF and IMP fixation values did not differ significantly during the acute and chronic period. However GBE administration produced a significant increase in rCBF on infarctus area and the rest of the brain and a moderate delayed IMP hyperfixation, + 6 %, on the cortical area of the same side and at the same level as the infarct.

In two patients treated with GBE from the first day of the infarct the discontinuation of the treatment for 5 days was accompanied by an increase in rCBF and delayed IMP hyperfixation (patients WEI.. and DEC..) (Table I). These two patients had a complete clinical recovery.

DISCUSSION

The interpretation of these preliminary results in the acute phase of infarct is difficult because spontaneous variations of CBF are expected according to the natural course of the disease and because patients are tested at different time intervals after the infarct. However, the results obtained on the first four patients on the 5th, 13th, 18th and 19th days are quite coherent whether GBE is administered from the very day of infarct or later. GBE administration induced a decrease of relative luxury perfusion in infarct areas as well as of rCBF around the infarct (penumbra ?) and the rest of the brain. Discontinuation of GBE was accompanied by a reappearance of the relative luxury perfusion. GBE also induced a decrease in the delayed IMP hyperfixation observed at 5 hours on the areas of relative luxury perfusion and discontinuation of GBE was followed by an increase in the delayed IMP hyperfixation.

The inversed rCBF response to GBE observed in one patient (GIR..) and the lack of response in another (LEC..) were not explained. Both presented a very small deep infarct and the accuracy of the measurement for such a small deep area could be questioned.

The real significance of the delayed IMP hyperfixation remains unknown. However it generally coincides with a relative luxury perfusion (9) and more accurately, a CT necrotic area, i.e. the ischemic core of the infarct. In the chronic period it is replaced by a large 5th hour hypofixation, which corresponds to the cyctic-like or sharp delineated hypodense area on the CT scan. Delayed IMP hyperfixation seems very interesting, since it provides evidence of a definitive necrotic area. It corresponded for 2 patients to a severe neurological deficit. Recently, among a larger group, 8 patients showed a delayed IMP hyperfixation, 6 maintened a severe neurological deficit. These results suggest that the presence of delayed IMP hyperfixation which seems indicative of necrotic, non-reversible lesions, should be used in making an early prognosis.

Interpretation of the delayed IMP hyperfixation is not clear. Although it generally corresponds to a relative luxury perfusion, it cannot be attributed to a "relative" high rCBF. In fact, the IMP brain curve which reaches a plateau at about 30 minutes for at least 24 hours is not due to an early fixation with perfect retention. It is rather the result of an ongoing process of accretion, or accumulation, and of a loss (1,7,10) and depends on CBF and cerebral metabolic activity. In the first 10 minutes the CBF component seems predominent and explains why it was proposed for CBF measurement. However, the discrepancies observed between rCBF and 10th minute IMP values in the first 4 patients of Table I indicated that a parenchymatous fixation, reflecting the cerebral metabolic activity, is already present. At 2 hours and particulary at 5 hours the metabolic activity component is predominent, indeed almost exclusive (7). At 2 and 5 hours IMP is fixed on cellular receptors and its distribution could be considered as a cellular function index. A hyperfixation at this period should be interpreted as cellular accumulation exceeding the loss. It could be due to very low cellular metabolism with slow IMP turnover and a low washout rate (7,8). The constant accumulation of IMP in these damaged areas could be related to the lipophily of the molecule and modifications of the local tissular pH.

The decrease in hyperfixation cannot be attributed only to an rCBF decrease because for the first patient of Table I, who was seen again 4 weeks later (41th day), the rCBF in the relative luxury perfusion area decreased from 45 to 36 while the hyperfixation remained practically unchanged with + 5 % and + 6 %. This suggests an effect on local metabolic activity and/or blood brain barrier permeability (7).

According to P. BRAQUET and al. (5), GBE is a free radical scavenger able to inhibit production and action of free radicals on cell-membrane components and the vascular system. Endothelium and B.B.B. hypermeability could so be reduced. GBE should exert a protective effect on hypoxic metabolic disorders and diminish cerebral oedema, especially its extra-cellular component.

REFERENCES

1. HILL T.C., HOLMAN B.L., LOVETT R., O'LEARY D.H., FRONT D., MAGISTRETTI P., ZIMMERMAN R.E., MOORE S., CLOUSE M.E., WU J.L., LIN T.H. and BALDWIN R.M. (1982) : Initial experience with SPECT (Single-Photon Computerized Tomography) of the brain using N-isopropyl I-123 p-iodoamphetamine : concise communication. J. Nucl. Med., 23, 191-195.
2. KUHL D.E., BARRIO J.R., HUANG S-C., SELIN C., ACKERMANN R.F., LEAR J.L., LIN T.H. and PHELPS M.E. (1982) : Quantifying local cerebral blood flow by N-Isopropyl-p-(123-I) Iodoamphetamine (IMP) tomographie. J. Nucl. Med., 23, 196-203.
3. LASSEN N.A., HENRIKSEN L., HOLM S., BARRY D.I., PAULSON O.B., VORSTRUP S., RAPIN J., PONCIN-LAFITTE M., MORETTI J.L., ASKIENAZY S. and RAYNAUD C. (1983) : Cerebral blood flow tomography : Xenon-133 compared with isopropyl-amphetamine-iodine-123 : concise communication. J. Nucl. Med., 24, 17-21.
4. LASSEN N.A. and ROMMER P. (1984) : Cerebral blood flow tomography by single photon emission computerized tomography using Xenon-133 in acute stroke. Drugs and Diseases. ed LASSEN N.A., HOSSMANN K.A., REIVICH M., AGNOLI A., CAHN J., J. John Libbey (eurotext) Paris-London, pp 125-134.
5. ETIENNE A., CHAPELAT M.Y., BRAQUET M., CLOSTRE F., DRIEU K., DeFEUDIS P.V. and BRAQUET P. (1984) : In vivo studies of free radical scavenging activity : relation to cerebral ischaemia. Cerebral Ischemia. ed. BES A., BRAQUET P., PAOLETTI R., SIESJO B., Excerpa Medica, Amsterdam, pp 379-383.
6. RAYNAUD C., RANCUREL G., SAMSON Y., KIEFFER E., BARON J.C., CABANIE E., LASSEN N.A., COMAR D., RAPIN J., MORETTI J.L., ASKIÉNAZY S., RICHARD S. and BOURDOISEAU M. (1984) : Measurement of oxygen metabolisme, 123-I-IMP distribution and rCBF-Xe in patients with ischemic vascular disease. Proc. Of 31. Annual Meeting os Soc. Nucl. Med., june, 1984? Los Angeles (Abstract). J. Nucl. Med., 25, p8.
7. RAYNAUD C., RANCUREL G., KIEFFET E., CABANIS E., BARON J.C., SAMSON Y., BOURDOISEAU M., BOURGUIGNON M., SYROTA A. and LASSEN N.A. (1985) : Chronic state after an ischemic stroke with persisting arterial occlusion : are low CBF/noram CT areas an expression of chronic penumbra or diaschisis ? In 4th Baune for on Conference, brain imaging. Ed CAHN J. and LASSEN N.A., published John Libbey, Eurotext, London.
8. RAYNAUD C., RANCUREL G., KIEFFER E., CABANIS E., LASSEN N.A., RICARD S., MORETTI J.L., BOURGUIGNON M., BOURDOISEAU M., ALLARD M. and SYROTA A. (1985) : Results obtained in recent infarcts with I-123 Isopropyl Iodoamphetamine (IMP) and Xe 133 with a high sensitivity spect system. Society of Nuclear Medecine 32nd Annual Meeting, Houston, Texas (June 1985) (abstract).
9. SKYHOJ OLSEN T. and LASSEN N.A. (1984) : A dynamic concept of middle cerebral artery occlusion and cerebral infarction in the acute state based on interpreting severe hyperemia as a sign of embolic migration. Stroke, 15, 458-468.
10. WINCHELL H.S., HORST W.D., BRAUN L., OLDENDORF W.H., HATTNER R. and PARKER H. (1980) : N-Isopropyl-(123-I)p-Iodoamphetamine : Single-pass brain uptake and washout : binding to brain synaptosomes : and localization in dog and monkey brain. J. Nucl. Med., 21, 947-952.